Metals in Surface Waters

Edited by
Herbert E. Allen
A. Wayne Garrison
George W. Luther III

Ann Arbor Press

Library of Congress Cataloging-in-Publication Data

Metals in surface waters / edited by Herbert E. Allen, A. Wayne Garrison and George W. Luther, III.

p. cm.

Includes bibliographical references (p.) and index.
IBSN 1-57504-087-5
1. Metals--Environmental aspects--Congresses. 2. Metals--Speciation--Congresses. 3. Water--Pollution--Congresses. I. Allen, Herbert E. (Herbert Ellis), 1939- . II. Garrison, A. Wayne. III. Luther, George W. IV. Metal Speciation and contamination of surface water.
TD427.M44M454 1997
628.1'683--dc21 97-33660
CIP

ANN ARBOR PRESS.
121 South Main Street, Chelsea, Michigan 48118
Ann Arbor Press is an imprint of Sleeping Bear Press

PRINTED IN THE UNITED STATES OF AMERICA

1 2 3 4 5 6 7 8 9 0

Metals in Surface Waters

PREFACE

Pollution is a design flaw, not a necessary correlation of production and commerce.

Surface waters, including streams, rivers, natural and man-made lakes and the oceans, are the support medium for most life on planet Earth. This life includes humans, who take most of their drinking water from surface systems. In addition, surface water bodies are also of great aesthetic and recreational value; many species of animals, including humans, find unparalleled sport and enjoyment in bathing, fishing and water play. In addition, the world's surface waters provide the great majority of water for manufacturing processes.

To our great misfortune, humans have allowed surface waters to be the prime repository of our wastes -- wastes from our bodies, our activities, and our great variety of conveniences and facilities, including manufacturing plants. These wastes include metals, some of which prove to be harmful to the very life that depends on the receiving water. The primary stress is toxicity to aquatic plant and animal organisms, but we are now very familiar with several secondary impacts; for example, bioaccumulation and bioconcentration of chemicals through the food chain that results in toxicity to non-aquatic species.

The chemical properties and behavior of these metal pollutants influence their fate, exposure and toxicity. The primary determinant of behavior is the chemical form in which the metal occurs -- referred to as the *species* of the metal. Metal *speciation*, then, is the process or combination of processes by which a metal arrives at the form(s) in which it is found in a particular state of the environment, often the equilibrium state. (Speciation can also rather loosely refer to the analytical determination of the species present in a particular state.) Valence changes of the metal atom, the formation of oxyanions, complexation with inorganic or organic ligands in solution, sorption to particulate or sedimentary matter, precipitation, and interaction with microbes are among the processes that lead to a new distribution of metal species.

This book is a compilation in chapter format of most of the oral presentations at the fifth Workshop on Metal Speciation, specifically entitled "Metal Speciation and Contamination of Surface Waters," which was held at Jekyll Island, Georgia, June 25th through the 28th, 1995. The previous Workshop compilations are listed below.

- *Metal Speciation: Theory, Analysis and Application* J.R. Kramer and H.E. Allen, Eds., Lewis Publishers, (Chelsea, MI 1988).
- *Metals in Groundwater*, H.E. Allen, E.M. Perdue and D.S. Brown, Eds., Lewis Publishers, (Chelsea, MI 1993).
- *Metal Speciation and Contamination of Soil*, H.E. Allen, C.P. Huang, G.W. Bailey and A.R. Bowers, Eds., Lewis Publishers, (Boca Raton, FL 1994).
- *Metal Contaminated Aquatic Sediments*, H.E. Allen, Ed., Ann Arbor Press, (Ann Arbor, MI 1995).

The objective of the workshop, so of this book, was to define the state of knowledge of metal contamination and speciation in surface waters, with the concomitant goal of defining related research needs. To do this, speakers from several disciplines, including analytical chemistry, aquatic chemistry, environmental engineering, oceanography and aquatic toxicology, were assembled to provide a broad and diverse program.

These chapters are arranged into four categories. The first category is *sources of metals to surface waters.* In **Chapter 1**, Thomas Church and Joseph Scudlark discuss the occurrence, sources and fate of trace metals in estuaries, using Delaware Bay as an example. Mark Wiesner and coworkers, in **Chapter 2**, examine the role of particles in urban runoff in transporting metals, with consideration of particle size distribution. Then in **Chapter 3**, James Patterson and coworkers discuss industrial discharges of metals to water, considering sources, concentrations, treatment technologies, and pollution prevention.

The next two chapters deal with the development of *background information necessary* for the *regulation of metal discharges.* Lial Tischler, **Chapter 4**, explores approaches to waste load allocations for metals, discussing modeling approaches and giving several case histories. In **Chapter 5**, Felix Locicero and coworkers discuss the development of site-specific copper criteria for the New York/New Jersey Harbor; this work includes a considerable amount of toxicity data.

The third category of chapters deal with *analytical methods for metals.* In **Chapter 6**, Carlton Hunt discusses the importance of "clean" metal sampling and analysis and shows that routine use of clean metals methods is necessary for an adequate scientific basis in many regulatory and assessment cases and for scientific investigation.

Constant van den Berg, in **Chapter 7**, describes his research on the use of cathodic stripping voltammetry to determine organic metal complexes and other species in seawater. In **Chapter 8**, Joseph Caruso and coworkers summarize their recent work using plasma mass spectrometric detection for metal species separated by various chromatographic techniques.

A fourth category of chapters relates to the *binding of metals to heterogeneous surfaces and dissolved humic substances.* Michael Perdue, in **Chapter 9**, discusses the interaction of trace metals with dissolved humic substances and his efforts to develop chemical speciation models to describe metal binding with humics. **Chapter 10**, by John Westall and coworkers, deals with modeling metal speciation in aquatic systems. Specifically, the authors review the development of models for the binding of metals to heterogeneous environmental materials and describe some new modeling approaches. In **Chapter 11**, Laura Sigg shows the relationship between field observations of partitioning between metals and suspended particles, including algae, and the surface chemical properties of the particles.

Chapter 12 is on the redox and pH chemistry of chromium in water and sediment. Here, William Patrick and coworkers, discuss the role of pH and redox conditions on the speciation and solubility of chromium in the aquatic environment, with special emphasis on wetlands.

We offer this book with the trust that its contents will provide the reader with a more comprehensive understanding of the aquatic chemistry of metals in surface waters and related sediments, more insight into the necessary analytical chemistry, and an increased knowledge of the sources of metals found in surface waters and the types of data necessary for metals regulation.

We thank all authors for their contributions and their cooperation during preparation of this book. The expert technical assistance of Ms. Dana M. Crumety is gratefully acknowledged. Finally, we thank the U.S. Environmental Protection Agency for its financial support.

Herbert E. Allen is Professor of Civil and Environmental Engineering at the University of Delaware, Newark, Delaware, U.S.A. Dr. Allen received his Ph.D. in Environmental Health Chemistry from the University of Michigan in 1974, his M.S. in Analytical Chemistry from Wayne State University in 1967, and his B.S. in Chemistry from the University of Michigan in 1962. He served on the faculty of the Department of Environmental Engineering at the Illinois Institute of Technology from 1974 to 1983. From 1983 to 1989 he was Professor of Chemistry and Director of the Environmental Studies Institute at Drexel University.

Dr. Allen has published more than 130 papers and chapters in books. His research interests concern the chemistry of trace metals and organics in contaminated and natural environments. He conducted research directed toward the development of standards for metals in soil, sediment and water that take into account metal speciation and bioavailability.

Dr. Allen is past-chairman of the Division of Environmental Chemistry of the American Chemical Society. He has been a frequent advisor to the World Health Organization, the Environmental Protection Agency and industry.

Arthur W. Garrison is a Research Chemist in the Exposure Research Division of the National Exposure Research Laboratory of the U.S. Environmental Protection Agency in Athens, Georgia.

Dr. Garrison received his Ph.D. in Organic Chemistry from Emory University in 1966, his M.S. in Chemistry from Clemson University in 1958, and his B.S. in Chemistry from The Citadel in 1956. Other than serving two years in the U.S. Army, his entire career has been spent with the U.S. EPA or its predecessor organizations. He held administrative positions for 10 years, including Chief of the Chemistry Branch at the Athens laboratory for 8 years.

In late 1992, Dr. Garrison was detailed to work for one year in the Institute for Ecological Chemistry of the German government's GBF research facility near Munich. There he pioneered in the applications of capillary electrophoresis to environmental problems, including the examination of humic substances and their interactions with pollutants and the separation and fate of the enantiomers of chiral organic pollutants. Most of his current research relates to chiral chemistry in the environment, and involves investigations of the degradation of chiral organic pollutants and their enantiomers. He also continues research on the character and reactions of humic substances through cooperation with his coworkers in Munich. Dr. Garrison has also been involved in research on the environmental chemistry of metals, including sorption of metal cations and the speciation of chromium species. He served several years on the EPA's Office of Water advisory board for development of sediment quality criteria for metals.

George W. Luther, III is Professor of Marine Chemistry in the College of Marine Studies at the University of Delaware, Lewes, Delaware, U.S.A. Dr. Luther received his Ph.D. in Physical-Inorganic Chemistry from the University of Pittsburgh in 1972 and his B.A. in Chemistry from LaSalle College in 1968. He served on the faculty of Kean College of New Jersey from 1972-1986 and was chairman of the Chemistry-Physics department from 1976-1984.

Dr. Luther has published more than 100 papers and chapters in books. His research interests concern the chemistry of non-metals such as sulfur and iodine and the chemistry of trace metals in natural environments with emphasis on redox speciation as well as organic and inorganic complexation of metals. He pioneered the application of frontier molecular orbital theory to better understand geochemical redox processes.

Dr. Luther is chairman of the Division of Geochemistry of the American Chemical Society and past-program chairman of the Division. He was the Gordon Research Conference chairman of Chemical Oceanography in 1995. He is a member of the editorial boards of *Aquatic Geochemistry* and *Marine Chemistry*.

CONTENTS

1

TRACE METALS IN ESTUARIES: A DELAWARE BAY SYNTHESIS

2

METALS AND COLLOIDS IN URBAN RUNOFF

3

INDUSTRIAL DISCHARGES OF METALS TO WATER

James W. Patterson, Roger A. Minear, Eduardo Gasca and Chriso Petropoulou

4

WASTE LOAD ALLOCATIONS FOR METALS

Lial Tischler

5

DEVELOPMENT OF SITE-SPECIFIC MARINE COPPER CRITERIA FOR NEW YORK/NEW JERSEY HARBOR

6

ACCURACY IN ANALYSIS - IMPORTANCE OF CLEAN METAL SAMPLING AND ANALYSIS

7

DETERMINATION OF METAL SPECIATION IN SEAWATER USING CATHODIC STRIPPING VOLTAMMETRY

8

NEW DEPARTURES FOR ELEMENTAL SPECIATION STUDIES UTILIZING PLASMA MASS SPECTROMETRIC DETECTION

9

METAL BINDING BY HUMIC SUBSTANCES IN SURFACE WATERS - EXPERIMENTAL AND MODELING CONSTRAINTS

10

MODELING METAL SPECIATION IN AQUATIC SYSTEMS

11

PARTITIONING OF METALS TO SUSPENDED PARTICLES

12

THE REDOX-pH CHEMISTRY OF CHROMIUM IN WATER AND SEDIMENT

Ronald D. DeLaune, William H. Patrick, Jr. and Tingzong Guo

TRACE METALS IN ESTUARIES: A DELAWARE BAY SYNTHESIS

Thomas M. Church and Joseph R. Scudlark
Graduate College of Marine Studies
University of Delaware
Newark, DE 19716-3501

1. INTRODUCTION

Trace metals are elements which exhibit either nutrient or toxic qualities for natural waters. Estuarine trace metals have both natural and anthropogenic sources, and are delivered by either eolian, fluvial or tidal means. However, these sources are not distinct, as some atmospheric component into a watershed is included in the fluvial component. Once within the estuary, the ability of a trace metal to survive transport to the sea appears to depend fundamentally on its aquatic chemical properties that render Type A metals reactive to particles or Type B complexed with organic compounds in solution. The fate of the geochemically reactive trace elements (Type A) is largely to become incorporated in reducing inter- or sub-tidal sediments. However, the sedimented trace metals may subsequently be remobilized by the actions of digenesis or dredging. The fate of biochemically reactive trace elements (Type B) is to be complexed in estuarine waters and transported largely to coastal waters. These principles are illustrated by a synthesis of results on the Delaware Bay Estuary. However, the biogeochemical processes elucidated for the Delaware could be extrapolated to other estuaries to help predict to what extent they are sources or sinks to coastal and open ocean waters.

The objectives of this paper address the following questions: What are trace metals and why are they important in estuaries? What are their sources for estuaries? What biogeochemical processes control their behavior in estuaries? What is their fate within or without the estuary? These basic questions must be answered prior to asking whether trace metals pose environmental problems or risks, either for living resources or humans using estuaries. First we present a general introduction to trace metals and estuaries. Then we illustrate answers to the above questions in a synthesis of results already published for Delaware Bay. Finally these results might be summarized for trace metal sources, biogeochemical reactions, and removal for estuaries in general.

2. TRACE METALS

Certain elements are called trace metals, although some are already abundant in rocks (e.g., Al, Fe, Mn) while others (e.g., Cd, Cr, Cu, Pb, Hg and Zn) are rarer in rocks until concentrated, used and disposed of by human practices. Some (e.g., As, Se, and Sb) are referred to as metalloids because they have both metal and non-metal properties, but being toxic are generally considered like metals.

Trace metals can be grouped according to their fundamental chemical properties (Table 1). Type A metals tend to form cations (positively charged ions) which react with negatively charged ions such as oxygen and suspended particles and thus become enriched in soils or sediments. Type B metals tend more to form stronger complexes or compounds with organic matter and thus are more stable for transport through aquatic environments.

Table 1. Chemical properties of trace metals.

Type A	Type B
• Iron (Fe), manganese, (Mn) cobalt, (Co), chromium (Cr)	• Zinc (Zn), nickel (Ni), copper (Cu), cadmium (Cd), silver (Ag), lead (Pb), mercury (Hg)
• Form inorganic ions in seawater	• Form organic complexes in seawater
• Particle reactions (precipitation, adsorption, flocculation)	• Biological reactions (assimilation)
• Removal by sedimentation	• Removal by tidal export

Many trace metals (e.g., Fe, Mn, Cu, Co, Zn, and Se) are essential for all life within certain concentration limits. However, above required concentrations, many of these elements (e.g., Zn, Cu, and Se) can have adverse effects. The same thing may be said of trace elements that are not essential for life (e.g., Cd, Hg, and Pb) -- above minimal concentrations, these elements are toxic. Some trace metals can have toxic impacts on estuarine and coastal ecosystems. Trace metals are important in estuarine ecology because they can be both essential for life, yet harmful in unnaturally high amounts. Generally speaking, the quantity of some macro- or micro-nutrients (including certain trace elements) will always limit how much algal growth occurs in natural waters. Conversely, it is just as likely that toxic trace elements are also limiting plant growth in natural waters such as estuaries being impacted by man.

3. ESTUARIES

Estuaries are confined drowned river valleys or lagoons where there is a mixing of fresh and saline waters. The sources of freshwater and its associated trace metals should include atmospheric deposition into watersheds and associated runoff into rivers and groundwaters. Ocean waters which provide a saline source of trace metals circulated landward by tidal forces across relatively shallow coastal shelves where atmospheric deposition should also be important.

To facilitate commercial transportation, urban populations and their industrial activities are often located seaward of the fall line, the landward limit of tidal influence. Historically, these activities should have made coastal waters, and estuaries in particular, vulnerable to trace-metal contamination. Here also, dredging is often used to maintain navigational channels and could serve to remobilize previously sedimented particles.

Land clearing for agricultural and industrial activities and subsequent erosion have converted most estuaries from being clear to being turbid. Thus in modern times, estuarine waters could be sufficiently clouded with suspended particles so as to limit the penetration of sunlight and restrict plant life. This is especially true of the upper, brackish reaches of estuaries where large particle loads should be trapped and resuspended by the opposing hydrological forces of river currents and tides, forming a so-called turbidity maximum zone. The turbidity of this zone should afford an excellent environment for particulate geochemical reactions such as adsorption and flocculation to occur for Type A metals. Consequently, light limitation in the zone could also restrict biochemical reactions related to photosynthesis for Type B metals. However, as particles settle out downstream and photosynthesis is less light limited, the saltier estuarine reaches should become a more biochemical zone. This could have the effect of releasing natural organic compounds during associated growth, death, and decay of estuarine organisms which can complex Type B metals.

4. RESULTS

The remainder of this paper is a synthesis of results for the Delaware Estuary that illustrates concepts and processes on trace-metal behavior in estuaries. The source materials are publications of the authors and coworkers on trace metals conducted under the auspices of the Delaware Estuary Program.

The Delaware Estuary presents a good case study since it exhibits classic characteristics of a large, well mixed, meso-tidal estuary processing a single dominant riverine source. Figure 1 is a diagrammatic representation of the Delaware showing its features, trace-metal sources and sinks, and biogeochemical processes.

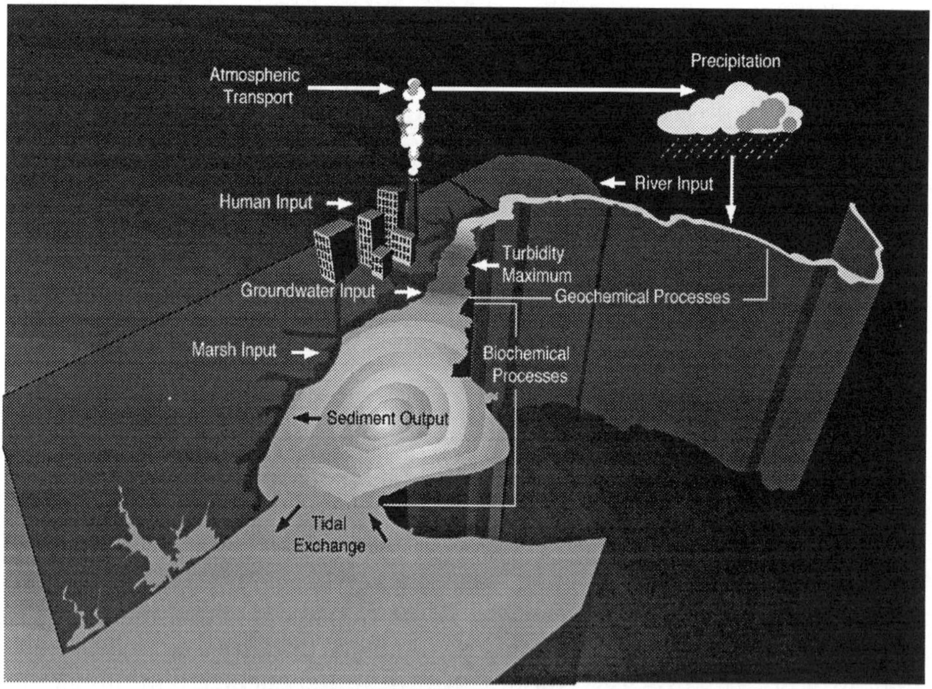

Figure 1. ***A Diagrammatic Representation of the Delaware Estuary as a Biogeochemical System. The processes that influence the behavior of trace metals in the estuary and the major inputs and outputs depicted would also be found in many estuaries with similar characteristics.***

5. TRACE-METAL INPUTS

The Delaware River is by far the dominant source of freshwater to the Delaware Estuary, providing nearly 80 percent of the gauged freshwater input. By the time the river reaches the fall line near Trenton, New Jersey, it carries a cumulative load of trace metals from atmospheric deposition, natural weathering reactions, and human-derived sources in the vast watershed upstream.

For the last decade, we have monitored trace metals in precipitation at Lewes, Delaware near the mouth of Delaware Bay [1,2]. The wet deposition is calculated as the product of the average concentration times the amount of precipitation. Dry deposition by aerosol is much more difficult to measure directly; depending on the metal, no more than half the total deposition is estimated to be dry based on assumed deposition velocities [3]. Trace-metal concentrations in Lewes precipitation appear to match or exceed those of other sites around the Chesapeake Bay region, suggesting regionally important sources or unique atmospheric scavenging characteristics in the coastal marine environment [3]. Based on similar studies for nitrogen, the precipitation data at Lewes should be representative of a homogeneous airshed for the Delaware watershed [4]. Thus, atmospheric trace-metal deposition at Lewes should be representative enough to be used to compare wet atmospheric inputs with riverine inputs [5].

Since metal concentrations in Delaware precipitation can exceed that of local river waters several fold [2], their contribution to the fluvial flux to the estuary could be significant. However, not all the trace metals deposited on the watershed reach the estuary through surface runoff or groundwater transport. This is illustrated in Figure 2, which shows the Delaware watershed and percentage of total atmospheric fluxes for several metals deposited in the watershed that reach the river at the fall line.

As Figure 2 shows, the atmospheric flux retained in the watershed from the atmosphere generally equals that entering the estuary [6]. This is because the area of the Delaware watershed exceeds by fifteen-fold that of its open water in the estuary. Cadmium is exceptional in this regard because as a large divalent cation (like Pb), it is soil humus reactive and retained, to a great degree, in the watershed. Thus, for all metals to a significant extent, the watershed soils are a large trace-metal reservoir that has been building up over the last two centuries of industrial activity. One exception is manganese that occurs as manganese oxide in soils and can be readily solubilized in groundwaters and transpired through leaf systems. The propensity of these trace-metal reserves to be remobilized at a later date depends both on the nature of the trace metals and the future use of the watershed and its associated groundwaters.

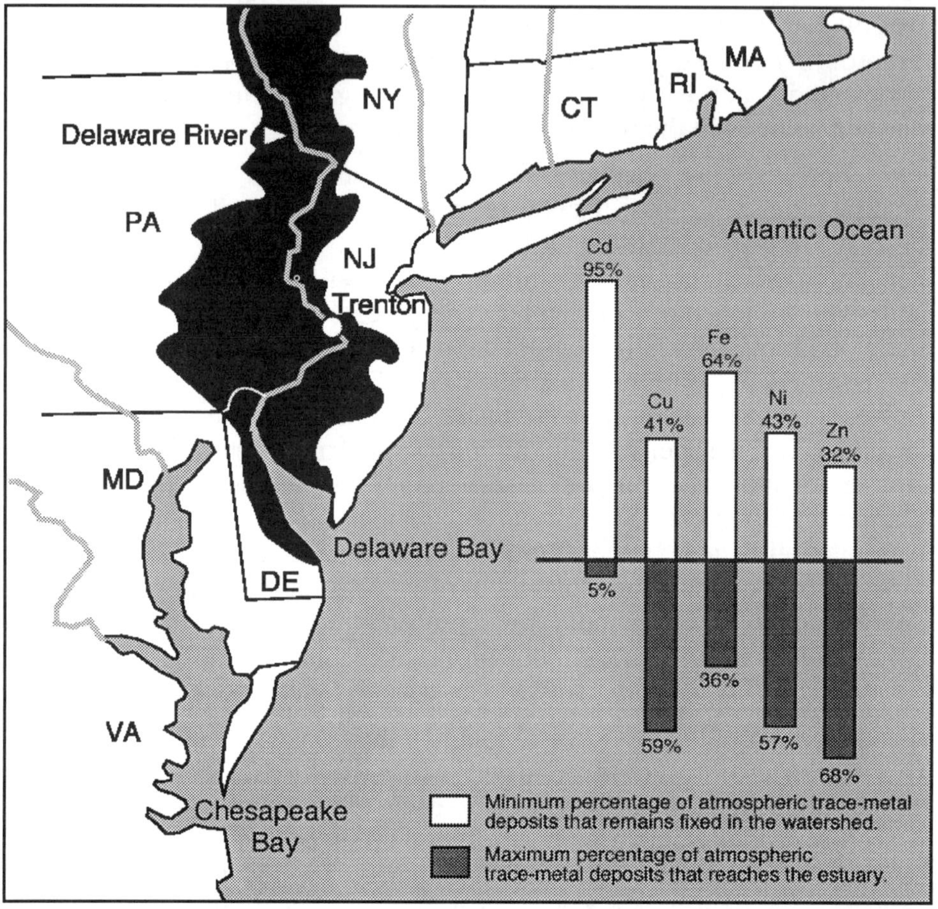

Figure 2. ***The Minimum Percentage of Total Atmospheric Trace-metal Flux Falling on the Delaware Estuary Watershed (map area shaded in black) that Remains in the Watershed versus the Maximum Percentage that Reaches the Estuary Across the Fall Line at Trenton, either Directly through Surface Runoff or Indirectly by Groundwater Seepage, for Several Metals.***

The metal burden of the Delaware River varies seasonally with minimum input occurring during the drier summer months and maximum loads during the wetter winter and spring. This most likely reflects the release of trace metals accumulated in the snow from atmospheric fallout and subsequent melting of this winter snowpack and the weathering of rocks, accelerated by associated freeze action. This is particularly evident for particle-reactive metals such as iron which enter the estuary in increased loads; greater than two-fold during the spring runoff. Other metals, such as zinc, show a much more uniform seasonal distribution; less than two-fold for reasons that will be subsequently discussed [6].

The impact of urban/industrial discharge can be seen by comparing concentrations of dissolved trace metals above and below the Philadelphia-Camden region of the upper Delaware Estuary where most increase several fold. When the flux of metals crossing the fall line north of Philadelphia is compared with that entering the estuary to the south, we find that 25-50 percent of the total non-crustal metal loads are introduced in this urban area alone [6]. Compared to other rivers on the East Coast, the Delaware is in the upper range for all metals and has the highest concentrations of some trace metals like zinc (Table 2), with levels similar to those of the Hudson near New York City.

Table 2. Concentration of dissolved trace metals in some east coast rivers in µg/L.

	Cd	Co	Cu	Fe	Mn	Ni	Pb	Zn	Ref.
Delaware	0.17	0.42	2.36	32.9	155	3.86	0.27	12.1	[6]
Susquehanna[1]	0.089	1.0	1.21	57.3	655	5.75	0.21	2.62	
Southeastern U. S. (avg.)	0.078		0.56	30.7	18	0.26		0.64	[7]
Hudson	0.25		3.24	31.9	10.7	2.41		8.83	[8]
Connecticut[2]	0.10		4.17	113	45.9			0.98	
Potomac								0.55	[9]

[1] R/V Cape Hatteras cruise March 3-10, 1982.
[2] D. Waslenchuck, personal communication.

6. TRACE-METAL BEHAVIOR IN THE DELAWARE ESTUARY

Plotting trace-metal concentrations *versus* salinity rather than distance downstream is more informative since salinity in the estuary increases predictably as one moves downstream. As illustrated in Figure 3, a conservative trace metal should exhibit a more or less linear concentration with salinity in the case of simple two-point mixing, while if a new source or sink of the metal is present downstream, the plot would be curvilinear.

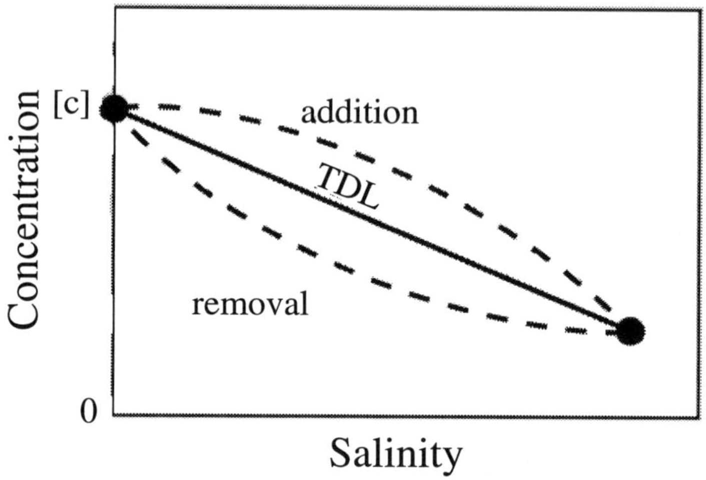

Figure 3. ***Theoretical Dilution Line (TDL) of an Ideal Dissolved Trace Metal as it Moves Downstream from Its Source [c]. Dashed lines indicate the effects of trace-metal addition or removal processes on the TDL.***

Samples for trace metals in the Delaware Estuary have been cleanly collected on 13 seasonal research cruises between 1978 and 1982 using GoFlo® bottles hung on Kevlar® line. The average distribution of seven trace metals in dissolved and particulate form as a function of salinity show distinct patterns according to metal type. Figures 4 and 5 present the data collected for manganese and copper as representative Type A and B metals, respectively [10,11].

As trace metals enter the Delaware Estuary, their biogeochemistry responds to estuarine particulate matter *via* three types of processes. One type is particle formation which encompasses the physicochemical processes that occur with the mixing of fresh and salt waters, such as adsorption and "flocculation" whereby dissolved and colloidal trace metals are converted to larger particulate forms. For example, dissolved Type A metals like manganese appear to be almost completely removed at the lower salinities near the turbidity maximum as they are converted from a dissolved to a particulate state (Figure 4a). Nevertheless, there is a suggestion of some seasonal desorption in more saline waters which is unique for Mn. At the same time, there is a corresponding increase in the percentage of particulate Type A metals, followed by a gradual decline as the particles settle out (Figure 4b).

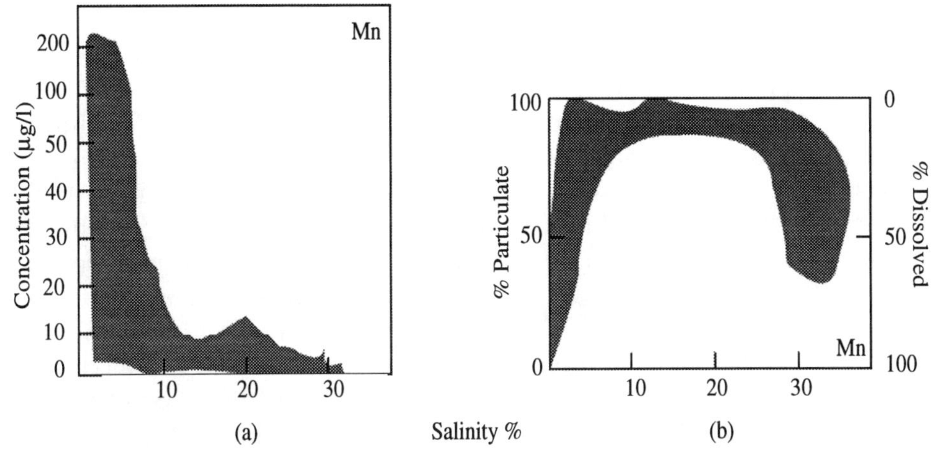

Figure 4. ***(a) Dissolved Concentration and (b) Percent Dissolved and Particulate Loadings versus Salinity for a Representative Type A Metal, Manganese (Mn). Note that salinity increases downstream in the estuary. The shaded areas envelope most data points measured on 13 seasonal research cruises between 1978 and 1982. The concave shape of the graphs reveals the nonconservative, particle-reactive behavior typical of Type A metals, resulting from the conversion of dissolved to particulate forms as salinity increases.***

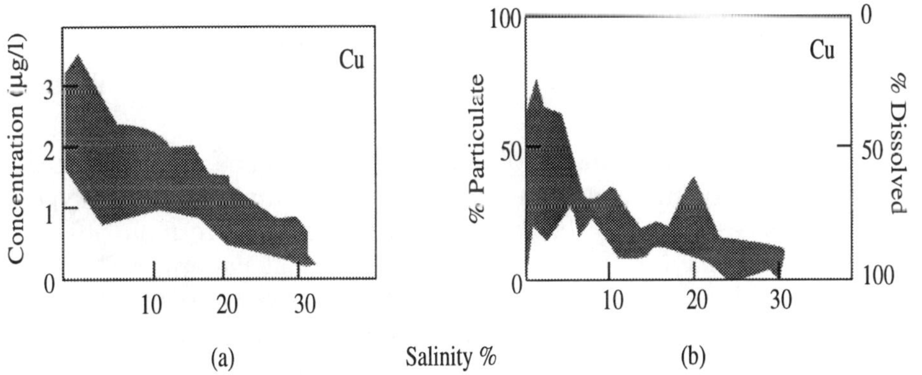

Figure 5. ***(a) Dissolved Concentration and (b) Percent Dissolved and Particulate Loadings*** versus ***Salinity for a Representative Type B Metal, Copper (Cu). The shaded areas envelope most data points measured on 13 seasonal research cruises between 1978 and 1982. The linear shape of the graphs reveals the conservative behavior typical of Type B metals, resulting from gradual dilution and biological processing downstream.***

A second process, physical resuspension, is a well demonstrated hydrologic feature that recurs in the turbidity maximum zone. The settling patterns occur between this point and the estuary mouth in what is termed the estuarine delta. The position and intensity of the Delaware turbidity maximum is seasonally controlled and migrates according to the balance of tidal and freshwater flows [12]. Heavier freshwater flows in the late winter and early spring both intensify the turbidity maximum and push it farther downstream. However, the flux of sediment crossing the turbidity maximum from upstream fluvial sources is too little to account for the flux of sediment settling in the estuarine delta and salt marshes [13]. In any case dredging removes the vast majority of upstream sediment which tends to settle into the navigation channels. One source of sediment may be derived from the net erosion and widening of Delaware Bay. However, to achieve closure on the sediment budget, the bulk of the sediment in the salt marshes and estuarine delta must be increasingly derived from offshore sources via winnowing of shelf sediments and shoreward tidal transport called residual circulation.

A third process of particulate trace-metal formation is biological. The high concentration of particles in the turbidity maximum tends to limit photosynthesis and the uptake of essential nutrient compounds. This allows for nutrient transmission to lower reaches of the estuary where they sustain primary production. Thus, secondary biological particles can be produced at higher-salinity reaches. These include the shells of phytoplankton such as diatoms that have assimilated trace metals in their structures or the particulate aggregates formed by deposit-feeding organisms. Both serve as reaction sites for trace-metal uptake [14,15].

However, these same biological processes can, in turn, provide the dissolved organic matter that forms stable complexes with Type B trace metals (such as Zn, Ni, Cu and Cd) which leads to their observed conservative behavior (Figure 5a). The seasonal patterns of Type B trace-metal behavior resemble many of the primary inorganic nutrients (nitrogen, phosphorus) suggesting their involvement with the biological productivity cycles of the estuary [16]. In fact, Cu itself can form robust complexes in the Delaware, presumably with organic material of biological origin [17]. With such biochemical processing, only the uncomplexed Type B metals can be actively assimilated and incorporated into soft tissues or hard shells, and thus the fate of these metals in estuaries depends on the fate of such parts. The soft tissue metals can degrade into dissolved complexes for export, while hard shells deposit into sediments which can subsequently dissolve.

In general, the concentrations of the particulate proportion of both types of trace metals decrease at high salinities (Figures 4 and 5). After forming either inorganic (Type A) or organic (Type B) particles, some of the estuarine metals settle out during the months of residence time in the Delaware Estuary in either the estuarine delta or the pervasive surrounding salt marshes. Here, in the sedimentary deposits, trace metals can be actively recycled as organic matter decomposes and the metals are converted to more soluble forms [18].

7. TRACE-METAL MASS BALANCE FOR THE DELAWARE ESTUARY

It is instructive to attempt to balance the input and output fluxes of trace metals in the Delaware Estuary because it gives an indication of the relative magnitude of the various sources and sinks. The comparison between the various inventories of trace metals reveals the retention of metals by sediments *versus* export to coastal waters. The trace-metal inputs, or sources, include riverine input, (dominated by the Delaware River), atmospheric deposition (both direct to open estuarine waters or indirect to the watershed), and groundwater (about which little is known). Other indirect sources include salt marshes and sediment exchange, representing the reprocessing of metals previously deposited as sediments. The primary trace-metal outputs, or sinks, are sedimentation and tidal flushing to coastal waters.

We have constructed a budget for total fluxes (F) of trace metals from data collected on the 13 seasonal cruises between 1978 and 1982 [10,19], which can be expressed by a steady state mass balance where r = riverine, m = marsh, a = atmosphere, x = tidal exchange, and s = sedimentation:

$$F_r + F_m + F_a = F_x + F_s \quad (1)$$

The left side of the equation gives inputs to the Delaware Estuary (dissolved plus acid leachable particulates [19]), while the right side yields outputs (Table 3).

7.1 Trace-Metal Inputs

The Delaware River is by far the largest freshwater contribution to the estuary (Figure 6). Thus, the freshwater discharge of the river times the concentration of the metal in the river gives the fluvial metal flux to the estuary (F_r). While other minor tributaries have been sampled on occasion, their metal concentrations are influenced primarily by tidal wetlands and, thus, their contributions to the riverine metal load are incorporated in the marsh term [6].

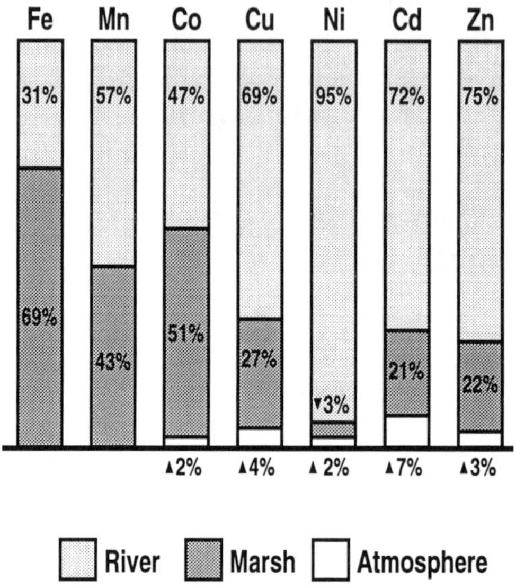

Figure 6. ***Percentage of Trace-metal Input to the Delaware Estuary from Each of the Three Major Sources for Various Metals.***

Table 3. Budget for the total fluxes of trace metals through the Delaware Estuary in g/s.

	Fe	Mn	Co	Cu	Ni	Cd	Zn
Input fluxes							
River	400.0	135.8	0.62	2.41	3.90	0.20	12.87
Marsh	885.0	102.0	0.67	0.96	0.14	0.06	2.80
Atmosphere	1.4	0.2	0.02	0.14	0.10	0.02	0.56
Total input flux	1286.2	238.0	1.31	3.51	4.14	0.28	17.23
Export fluxes							
Tidal exchange	176.4	19.0	0.172	2.21	2.67	0.191	7.48
Sediments *(calculated)*	1110.0	219.0	1.14	1.30	1.47	0.09	9.75

The flux of ungauged groundwater is estimated to equal the gauged freshwater inputs [10], but without sufficient trace-metal data, their metal contribution to the estuary is difficult to quantitatively assess. The groundwater terms could be significant for those metals (e.g., Fe and Mn) that can be naturally reduced and solubilized under low-oxygen subterranean conditions. In any case the fluvial metal load measured contains unresolved groundwater, particularly during base flow conditions of drier seasons which is mostly groundwater [6].

Atmospheric deposition of metals directly to the surface waters of the Delaware Estuary (F_a) does not appear to be a significant quantitative term (Table 3). However, this may largely be an underestimation for two reasons. First, the area of the Delaware Estuary's watershed is 15 times larger than that of the Bay itself; thus the actual atmospheric trace-metal component that crosses the fall line and, thus, is counted as "fluvial" appears to be significant (Figure 2). Second, direct atmospheric deposition to the estuarine surface waters could be qualitatively important to phytoplankton living in the euphotic zone. On the other hand, in estuaries with larger watersheds relative to open water but smaller fluvial/anthropogenic loading (e.g., Chesapeake Bay), the relative atmospheric input is more important [5].

Salt marshes are prominent features of the Delaware Estuary, and as they are continually undergoing active sulfur and metal cycling, their metal inputs may be important [18,20-23]. Further, salt marshes appear to be efficient collectors of atmospheric metal input, and even focus metal fallout from a much larger tidal area [24]. This is because the tidal waters bring trace metals into the salt marsh from the bay on a regular basis which, being enriched in the surface microlayer, can be scavenged by the marsh vegetation or surface sediments [19,25,26].

As such, the trace-metal contributions from Delaware salt marshes are secondary to the estuary since they come largely from atmospheric deposition and reflux of sedimented particles, and thus are already included in the primary atmospheric and fluvial sources of the mass balance. Nevertheless, for comparisons, the salt marsh metal source is obtained by multiplying the net tidal flushing of the salt-marsh creeks (obtained by salt balance) by the highest concentrations observed in the creeks. Although this may be an upper estimate, salt marsh metal contributions to the Delaware Estuary (F_m) appear in Table 3 to be equivalent to that of the river itself (F_r).

7.2 Trace-Metal Outputs

Trace-metal sinks from sedimentation (F_S) are difficult to estimate for reasons that were discussed for the salt marsh source. This is the uncertain heterogeneity of trace-metals in sediment regimes of the Delaware, and the associated trace metal diagenetic activity within its predominant salt marshes, the largest area of significant net sedimentation for the Delaware. Thus, we have chosen to use the tidal export approach and calculate the sediment flux inventory by difference. Consequently, F_S can be calculated by difference once the other term for estuarine export is quantified (F_X) with some uncertainty. Nevertheless, a direct estimate of F_S agrees within a factor of two for most metals, which supports this approach [10].

Quantifying total trace-metal export to or import from coastal waters is also difficult because the net flux of water by means of residual circulation is a small difference between two large tidal terms. Nevertheless, the net-water export can be estimated using a simple box model for the surface and bottom layers which is constrained by salt balance. Multiplying this net water flux by the concentration for lower estuary metals, yields the net metal export flux (F_X). This approach is superior to simply assuming the net tidal water export equals the original fluvial input because of the large amount of water evapotranspired by the extensive Delaware salt marshes [10].

The ability of trace metals to be removed or transit the estuary is the result of the mass balance (Figure 7). The Type A metals which are geochemically reactive are, to a very large degree, retained as sediments in the estuary. However, Type B metals, which are more biochemically active, tend to be sedimented and exported to coastal waters in equivalent amounts. This means that estuaries, in general, may be important sinks for Type A metals, yet significant sources of Type B metals for coastal shelf waters. Thus, to what extent Type B estuarine trace metals are essential or toxic for the health of estuarine biota is still unknown, but the ecosystems of estuarine and coastal waters should be fundamentally linked.

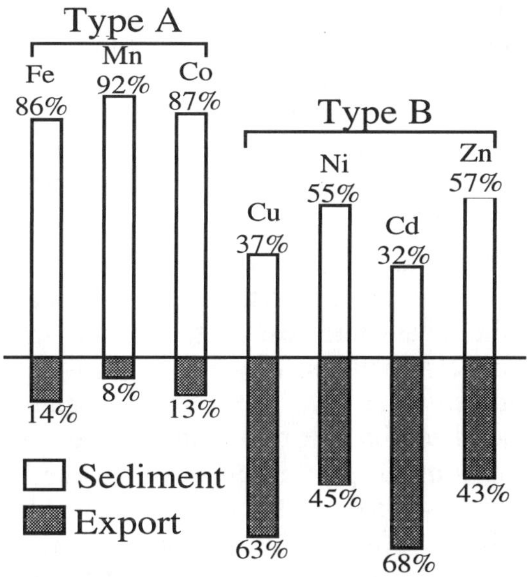

Figure 7. ***Percentage of Trace-metal Output from the Delaware Estuary* via *each of the Two Major Sinks for Various Metals.***

8. CONCLUSIONS

From the case study presented for the Delaware Estuary, a number of conclusions can be reached concerning the behavior of trace metals in estuaries. Although these findings are specific to the Delaware, the concepts should generally apply to other large estuaries with similar characteristics. The comparison in trace-metal behavior between estuaries depends as much on differences in their physical characteristics as well as common metal sources and fundamental biogeochemistry. Though beyond the scope of this paper, this could be borne out in similar studies of trace metals in other estuaries and comparison with the data presented here for the Delaware.

8.1 Trace Metal Sources to Estuaries

The sources of trace metals to estuaries, while still in part natural, have become more dominated by human activity. First, while the natural weathering of rocks has introduced trace metals to estuaries over geologic time, these weathering reactions can be disturbed by environmental degradation. Foremost, the fallout of atmospheric pollutants can fundamentally alter or accelerate chemical weathering by the increased acidity of precipitation downwind of industrial areas. However, the atmosphere itself can also contribute large amounts of toxic trace metals which are deposited by both dry aerosol fallout and wet scavenging as precipitation into watersheds. This is because metals originate concurrently with acids in the burning of fossil fuels for power or during metallurgical processing.

Second, precipitation runoff contains trace metals and finds its way into aquifers, where it can be retained or redischarged according to the basic hydraulic characteristics of the watershed and associated subterranean chemical reactions as well. Third, there are recycling processes wherein estuarine trace metals in sedimented particles can be subsequently remobilized back to the water. These processes include the action of salt ions displacing metals adsorbed to particles, or diagenetic reactions associated with organic matter decomposition that increase the solubility of many trace metals through reduction in complexation. Fourth, the sea itself may introduce trace metals to estuaries which are accumulated from atmospheric fallout as well as from upwelling, or sediment digenesis over the large area of coastal shelves, and carried landward with the tide.

8.2 Trace Metal Biogeochemical Reactions in Estuaries

Upon reaching the estuarine environment, trace metals display a variety of behaviors. Once encountering the first traces of sea salt, most Type A trace metals carried by rivers are converted from dissolved to particulate forms through the processes of precipitation, adsorption, or "flocculation" (whereby dissolved or smaller colloidal particles aggregate to form larger particles). However, with increasing salinity, some weakly adsorbed trace metals (e.g., Mn, Zn, Cd) can be desorbed through the action of ion exchange.

Trace metals also play several unique roles in estuaries in their biochemistry. Some become involved in primary production by algae, which can result in uptake and recycling of trace metals along with organic matter. Trace metals can react directly with nutrients like phosphate, or reduced species like sulfide by forming solid precipitates that are buried in the sediments. Some trace metals involved in sedimentary reactions can provide electron transfer for bacterial activity, degrading organic matter in the sediments, and thus, contribute directly to the cycling of both trace metals and nutrients.

8.3 Trace Metal Removal from Estuaries

As a result of their behavior, trace metals have a number of removal pathways in the estuary. Type A trace metals, which readily flocculate from dissolved to particulate form, can settle out as integral components of the bottom sediments. The larger particles settle out to subtidal areas. However, most of these sediments are extensively dredged to maintain shipping, and are removed from the system to dry land. The finer, flocculated precipitates are often organic rich and may remain suspended long enough to be transported to more remote settling areas such as surrounding salt marshes or exported to offshore coastal areas.

After deposition in estuarine sediments, organic matter degradation by bacteria can release particulate trace metals (through metal reduction and decomposition of organo-metallic complexes), resulting either in their dissolution and return to the water column or in the formation of new solid phases. This rerelease of metals from the sediment to the water *via* sediment digenesis, is largely promoted by the sulfate ion in sea water, the primary oxidizing agent. Since a primary by-product of sulfate reduction is sulfide, many trace metals in estuarine sediments are finally converted to more insoluble sulfide precipitates, helping to sequester them from saline waters unless dredged or seasonally oxidized in salt marshes by the action of rooted vegetation or sediment desiccation.

Trace metals may also be removed from estuaries through uptake by organisms and incorporated into the marine food web which may eventually lead to human consumption. In addition to absorbing or removing dissolved trace metals from the water, many aquatic organisms can actively excrete organic compounds called chelators that bind trace metals, converting them to stable dissolved forms that can be less readily incorporated into living tissues. However, estuarine organisms are usually viewed as merely a temporary repository for trace metals since the organisms will eventually die and be deposited in the sediments or washed to sea like other forms of trace metals. Ultimately, estuarine trace metals have just two major fates: either to be sequestered as particles and remain in estuarine sediments, or to be exported out of an estuary largely in dissolved form to offshore waters.

Acknowledgments: This research was supported by grants from the National Oceanic and Atmospheric Administration as part of the University of Delaware Sea Grant College Program. Much of the research was part of the Delaware Estuary Program with additional support from the Delaware River and Bay Authority. Other funding was obtained from the Ocean Sciences Division of the National Science Foundation.

The authors wish to acknowledge the many talented marine scientists (J. Tramontano, S. Murray) and graduate students (R. Pellenbarg, C. Lord, K. Eastman, K. Johnston, and T. Ferdelman) who over the years have been part of this research. Their fine efforts have contributed largely to the results presented in this synopsis.

In addition to one anonymous reviewer, the authors also wish to thank the staff members of the Marine Communications Office for their assistance in producing this document: Elizabeth Chajes, Editor; David Barczak, Art Director; and Pamela Donnelly, Production Manager.

REFERENCES

1. Church, T.M., J.M. Tramontano, J.R. Scudlark, T.D. Jickells, J. J. Tokos, Jr., A.H. Knap and J.N. Galloway, "The Wet Deposition of Trace Metals to the Western Atlantic Ocean at the Mid-Atlantic Coast and on Bermuda," *Atmos. Environ.* 18:2657-2664 (1984).

2. Church, T.M. and J.R. Scudlark, "Trace Elements in Precipitation at the Coast: A Successful Record Since 1982," in USDA Forest Service Report NC-150, *Symposium Proceedings, Philadelphia, PA, October 8, 1991*, pp. 45-46 (1992).

3. Scudlark, J.R., K.M. Conko and T.M. Church, "Atmospheric Wet Deposition of Trace Metals to Chesapeake Bay: CBAD Study Year 1 Results," *Atmos. Environ.* 28:1487-1489 (1994).

4. Scudlark, J.R. and T.M. Church, "Atmospheric Input of Inorganic Nitrogen to Delaware Bay," *Estuaries* 16:747-759 (1993).

5. Scudlark, J.R. and T.M. Church, "Atmospheric Deposition of Trace Elements to the Mid-Atlantic Bight," in *Atmospheric Deposition of Contaminants to the Great Lakes and Coastal Waters*, J.L. Baker, Ed., SETAC Press (Pensacola, FL 1997).

6. Church, T.M., J.M. Tramontano, J.R. Scudlark and S.L. Murray, "Trace Metals in the Waters of the Delaware Estuary," in *Ecology and Restoration of the Delaware River Basin,* S.K. Majumdar, E.W. Miller and L.E. Sage, Eds., pp. 93-115. Academy of Natural Sciences (Philadelphia, PA 1988).

7. Windom, H.L. and R.G. Smith, "Factors Influencing the Concentration and Distribution of Trace Metals in the South Atlantic Bight," in *Oceanography of the Southeastern United States Continental Shelf* (Coastal Estuarine Science 2). L.P. Atkinson, D.W. Menzel and K.A. Bush, Eds., pp. 141-152. American Geophysical Union (Washington, DC 1984).

8. Klinkhammer, G.P. and M.L. Bender, "Trace Metal Distribution in the Hudson River Estuary," *Est. Coast. Shelf Sci.* 12:629-643 (1981).

9. Shiller, A.M. and E. Boyle, "Dissolved Zinc in Rivers," *Nature* 317:49-52 (1985).

10. Church, T.M., "Biogeochemical Factors Influencing the Residence Time of Microconstituents in a Large Tidal Estuary, Delaware Bay," *Mar. Chem.* 18:393-406 (1986).

11. Church, T.M., J.M. Tramontano and S. Murray, "Trace Metal Fluxes Through the Delaware Estuary,"in Proc. of Int'l. Council for Exploration of the Sea (ICES) Symp. Contaminant Fluxes through the Coastal Zone, Nantes, France, May 1984. *Rapp. P.-v Reun. Cons. int. Explor. Merc* 186:271-277 (1986).

12. Biggs, R.B., J.H. Sharp, T.M. Church and J.M. Tramontano, "Optical Properties, Suspended Sediments, and Chemistry Associated with the Turbidity Maxima of the Delaware Estuary," *Can. J. Fish Aquatic Sci.* 4:172-179 (1983).

13. Biggs, R.B. and E.L. Beasley, "Bottom and Suspended Sediments in the Delaware River and Estuary," in *Ecology and Restoration of the Delaware River Basin*, S.K. Majumdar, E.W. Miller and L.E. Sage Eds., pp. 116-131. Pennsylvania Academy of Science (Philadelphia, PA 1988).

14. Sharp, J.H., The Delaware Estuary: Research as Background for Estuarine Management and Development. Report to Delaware River Bay Authority. University of Delaware Graduate College of Marine Studies and New Jersey Marine Sciences Consortium (1983).

15. Pennock, J., "Chlorophyll Distributions in the Delaware Estuary: Regulation by Light Limitation," *Est. Coast. Shelf Sci.* 21:711-725 (1985).

16. Sharp, J.H., J.R. Pennock, T.M. Church, J.M. Tramontano and L.A. Cifuentes, "The Estuarine Interaction of Nutrients, Organics, and Metals: A Case Study in the Delaware Estuary," in *The Estuary as a Filter*, V.S. Kennedy, Ed., pp. 241-257. Academic Press (New York, NY 1984).

17. Johnston, K., "Estuarine Organic Complexation of Copper in the Delaware Bay," Masters Thesis, University of Delaware, College of Marine Studies, 1989.

18. Lord, C.J. and T.M. Church, "The Geochemistry of Salt Marshes: Sedimentary Ion Diffusion, Sulfate Reduction, and Pyritization," *Geochim. Cosmochim. Acta.*47:1381-1391 (1983).

19. Church, T.M., M. Bernat and P. Sharma, "Distribution of Natural Uranium, Thorium, Lead, and Polonium Radionuclides in Tidal Phases of a Delaware Salt Marsh," *Estuaries* 9:2-8 (1986).

20. Bouleague, J., C.J. Lord and T.M. Church, "Sulfur Speciation and Associated Trace Metals (Fe, Cu) in the Pore Waters of Great Marsh, Delaware," *Geochim. Cosmochim. Acta* 46:453-464 (1982).

21. Eastman, K.W. and T.M. Church, "Behavior of Iron, Manganese, Phosphate and Humic Acid During Mixing in a Delaware Salt Marsh Creek," *Est. Coastal Shelf Sci.* 18:447-458 (1984).

22. Luther, III, G.W., T.M. Church, J.R. Scudlark and M. Cosman, "Inorganic and Organic Sulfur Cycling in Salt-Marsh Pore Waters," *Science* 232:746-749 (1986).

23. Ferdelman. T.G., T.M. Church and G.W. Luther, III, "Sulfur Enrichment of Humic Substances in a Delaware Salt Marsh Sediment Core," *Geochim. Cosmochim. Acta* 55:979-988 (1991).

24. Church, T.M., C.J. Lord and B.L.K. Somayajulu, "Uranium, Thorium, and Lead Nuclides in a Delaware Salt Marsh Sediment," *Est. Coastal Shelf Sci.* 13:267-275 (1981).

25. Pellenbarg, R. and T.M. Church, "The Estuarine Surface Microlayer and Trace Metal Cycling in a Salt Marsh," *Science* 203:1010-1012 (1979).

26. Church, T.M., M.M. Sarin, M.Q. Fleisher and T.G. Ferdelman, "Salt Marshes: An Important Coastal Sink for Dissolved Uranium," *Geochim. Cosmochim. Acta* 60:3879-3887 (1981).

METALS AND COLLOIDS IN URBAN RUNOFF

Mark R. Wiesner, Gregory W. Characklis and Drahomira Brejchovà
Department of Environmental Science and Engineering
Rice University
Houston, TX 77251

1. INTRODUCTION

The environmental impact of urban areas is widely recognized. One undesirable result of anthropogenic activity is the transport of significant quantities of solids and metals to nearby surface waters. This paper examines the role of particles in urban runoff in transporting metals, and considers the implications of the particle size distribution in metal transport and transformation.

2. METALS IN URBAN RUNOFF

Most materials loading occurs as the result of storm events. In a recent study of non-point source loadings to Galveston Bay in Texas, it was estimated that under average conditions, over 40% of the suspended solids entering the bay originate from runoff from the Houston Metropolitan Area [1]. Sartor and Boyd [2,3] concluded that, in a hypothetical city of 100,000 people, runoff from street surfaces alone could provide a significant source of contaminants to receiving waters. Furthermore, they found that the smallest fraction of particles considered in their work (<43μm) contained only 5.9% of the total mass of particles but over 50% of the heavy metals present. Zinc and lead were the most prevalent metals found amongst the street surface materials. Colston [4] concluded that urban runoff in Durham, NC overshadowed sewage effluent as the primary detriment to receiving water quality during wet weather periods (~20% of the year). While urban runoff contained only half the yearly output of organic waste found in municipal sewage, it carried as much as fifty times the annual loadings of suspended solids and heavy metals. Newton *et al.* [5] were puzzled by excessive amounts of lead present in an Oklahoma River that lacked any discernible pollution source. Their research eventually linked the lead to vehicle emissions that had been deposited on road surfaces and swept into receiving waters by stormwater runoff. Hedley and Lockley [6] undertook a comprehensive study of road surface runoff and reported evidence of a direct relationship between suspended solids and heavy metals. Once again, the highest metal loadings came from zinc and lead. Hunter and coworkers [7] observed that approximately 50% of the heavy metals were associated with particles in one urban watershed. Analysis of the metals concentration in precipitation indicated that rainwater itself was responsible for only 2-6% of the heavy metals in this stormwater.

Wilber and Hunter [8] compared the metals contributions to contamination of a New Jersey River from stormwater runoff and secondary sewage treatment effluent as well as rainwater. They concluded that stormwater was responsible for as much as 86% of the total metal loading to the river and that 84% of the total metals were either zinc or lead. In a later study [9], they observed that in small watersheds first flush events accounted for loadings as much as two orders of magnitude larger than the average loading. In yet another paper, Whipple *et al.* [10] challenged a common assumption that pollutant loadings in runoff were directly tied to the length of time between street flushing. Their data suggested little correlation between storm frequency and pollutant loading. They found that in most cases, regardless of the antecedent dry period, pollutant loadings were directly linked to total volume of stormwater runoff. In later work [11], the same Rutgers research group examined stormwater runoff from specific land use areas and determined that, while loadings of zinc and chromium per square mile increased for multiple-family housing, the values for lead and nickel dropped and loadings of copper remained roughly the same.

Anthropogenic sources of metals in urban runoff may include the combustion of materials and their subsequent atmospheric deposition, corrosion products, unmonitored industrial discharges, contaminated sites, and tire wear. There is considerable variability in the concentrations of metals measured on the particles in urban runoff and on impacted sediments (Table 1).

Table 1. Concentrations of metals (μg/g) on particles in urban runoff and recently deposited sediments as reported for several urban areas.

Location	Pb	Zn	Cd	Cu	Cr	Ni	Fe	Ba	Ref.
Valladolid, Spain	18.77	245.49	1.05	66.53					[12]
San Jose, Costa Rica	64	310		550	660				[13]
Venice, Italy		163					17603		[14]
Rouen, France		550					28345		[15]
Houston, Texas (this work)	100	4000		150			10000	1000	
Hong Kong	50.9	131	0.93	42.3	120				[16]
Hamilton, New Zealand*	970	1046.6		122.9					[17]
Tokyo, Japan		571			373	133	44928		[18]
Lodi, NJ	3990	1210		325	58	57	28400		[19]

* Assumes all metal associated with solids.

This variability may reflect changes in watersheds over time as well as the adoption of improved methodologies for measuring metals. For example, the concentration of lead in urban runoff has declined over the last 20 years in cities where the consumption of gasoline with lead additives has been reduced. In other cases, concentrations of mercury have decreased from levels violating regulatory standards to near detection, apparently due to the use of "clean" sampling techniques.

Zinc is a particularly interesting metal in urban runoff. It appears to exhibit a variability in particulate concentration that may be largely explained by differences in population density and motor vehicle use. Corrosion of galvanized metal and tire wear are frequently cited sources of zinc in urban runoff. A plot of the zinc concentration on particles and sediments versus fuel consumption reveals a direct correlation between increasing zinc concentration and fuel consumption (Figure 1). This suggests that tire wear may be the principal source of particulate zinc in urban runoff, and that much of the global variability in particulate zinc concentrations might be explained based on motor vehicle use. A similar plot of the concentrations of other metals data versus fuel consumption (or urban density) does not yield a discernible trend. However, were data on cadmium concentrations available, we would expect it to show a trend similar to that on zinc, as cadmium is an impurity commonly associated with zinc.

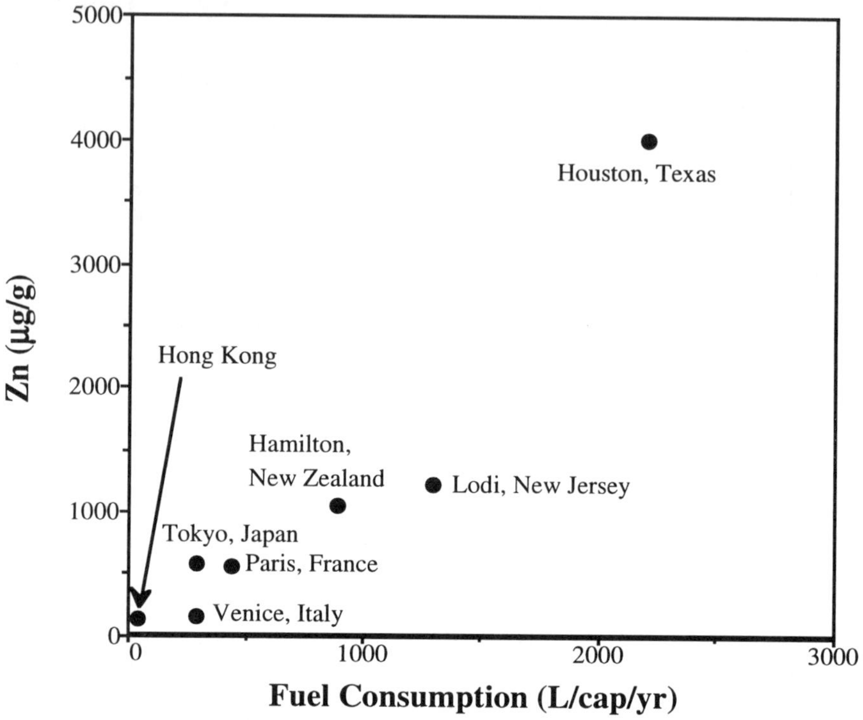

Figure 1. Zinc Concentrations on Particles and Sediments in Several Urban Areas as a Function of Estimated Per Capita Fuel Consumption.

3. PARTICLE SIZE AND METALS

Information on the size distribution of particles in urban runoff is surprisingly limited, despite its importance in characterizing the source and providing a strategy for control of pollution from runoff. An important exception is the 1979 work of Wilber and Hunter which examined the effects of urbanization on the distribution of heavy metals in the Saddle River sediments which were considered to be largely affected by urban runoff [19]. They fractionated sediment materials into particle size ranges varying from >2000 µm to approximately 0.01 µm and found increasing metals concentrations with decreasing particle size. The larger size fractions displayed successively higher metals concentrations as the river flowed through the New Jersey community of Lodi and became more susceptible to the effects of urban runoff. While the smaller size fractions continued to maintain higher metals concentrations as the river flowed through town, it was the fractions greater than 420 μm that manifested the largest increases in concentration. Particles in the smaller size fractions were thought to be more susceptible to scouring and transport than the larger particles. They also reasoned that larger particles provided a better long-term record of urban activity than did smaller particles, and that small suspended particles probably remained in suspension until they reached an estuary or some other low shear environment. Wilber and Hunter noted that the samples obtained just downstream from the secondary treatment plant outfalls contained relatively large concentrations of metals in the smaller size fractions (<125 μm). They hypothesized that this may have been the result of the metals in the river combining with the colloidal organic matter released in the treatment plant effluent.

This hypothesis was supported by additional studies that concluded that as much as 30% of the most common metals, lead and zinc, were associated with organic materials, and that the fraction of metals complexed with organic materials in the sediment increased as the river came under increasing influence from the community. Analysis of settleable and nonsettleable fractions of zinc in urban runoff to this system indicated that zinc was about evenly split between these two fractions [20]. In apparent contrast with earlier work, the length of the dry period between storms was considered to be an important consideration in determining pollutant loadings from storms.

Tanizaki *et al.* considered the relationship between suspended particles, metals concentrations and particle size distributions of streams in metropolitan Tokyo [18]. Although their work did not specifically address urban runoff, the streams they sampled were under considerable influence from a large urban area. Samples were fractionated by means of parallel filtration into classes ranging from greater than 0.45 μm to the fraction passing through a 500 MW cut-off membrane. Both metals and total organic carbon (TOC) concentrations were measured in each fraction. They concluded that zinc (and many of the rare earth elements present at much lower concentrations) were very likely to form complexes with naturally occurring organic ligands in the dissolved phase as well as with particulate-phase organic materials.

In recent work, the majority of particles in the urban runoff from a Houston watershed were found to reside in the size fraction below 2 μm [21]. This was true during periods of background flow as well as during storm events (Figure 2).

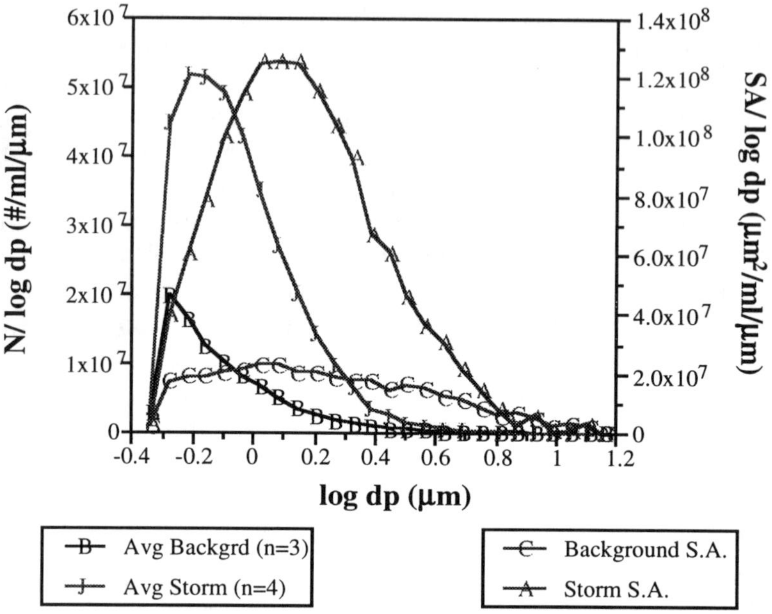

Figure 2. Particle Number and Surface Area (S.A.) Distributions Averaged Over Several Grab Samples During Background and Storm Flow Conditions on Brays Bayou at Gessner.

One implication of this observation is that the largest concentrations of particle surface area for the adsorption of heavy metals and other contaminants will be found on smaller particles that will tend to settle more slowly in the absence of any transformations. These particles may scavenge contaminants from the water column. The concentrations of some metals, such as iron in this instance, are more highly correlated with the number concentration of colloidal particles than with surface area. In many instances, it is the concentration of organic carbon, both particulate and dissolved, that correlates with metal concentrations. For example, during storm events, zinc in the Houston runoff water tended to display a bimodal distribution between dissolved and larger particulate fractions, while iron was found to reside almost entirely in the larger colloidal fraction. Similar trends were observed by Tanizaki and co-workers [18] in their measurements in urban waterways (Figure 3).

These differences in the apparent size fractionation of metals also appear to be reflected in differences in the times at which metals reached their peak concentrations during runoff events. The association of metals with particulate or organic phases entering the waterway determines when materials are "eluted" from the watershed. For instance, a trace of zinc concentrations closely followed that of TOC during the course of storm events we observed in Brays Bayou (Figure 4).

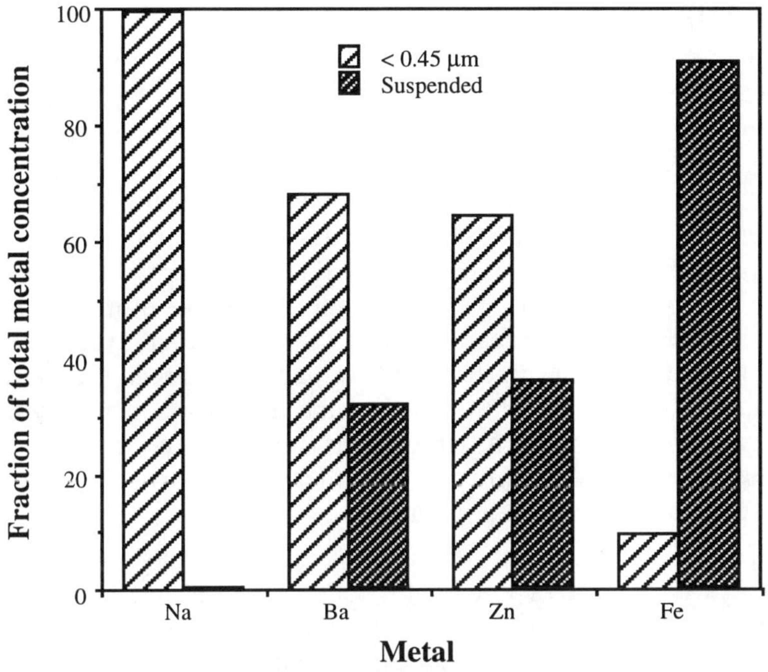

Figure 3. Size Distribution of Several Metals in an Urban Waterway. (Data taken from Tanizaki and coworkers [18]).

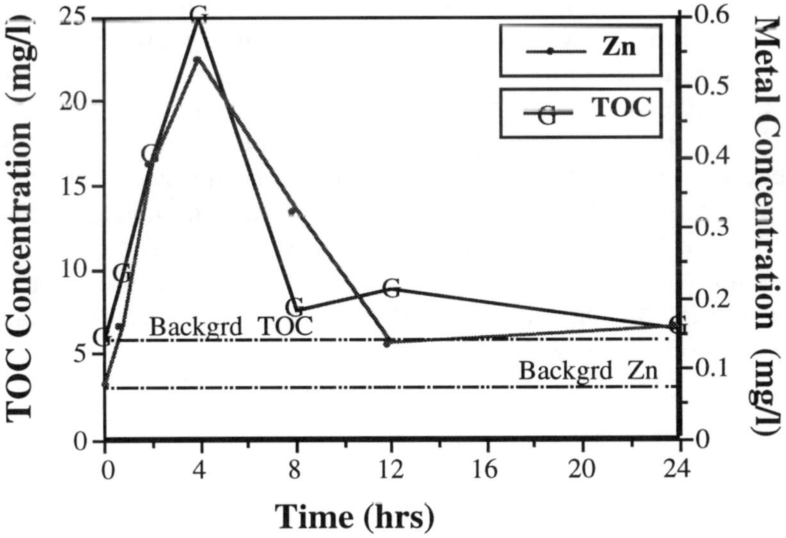

Figure 4. The Evolution of Zinc and Total Organic Carbon Concentrations During a Storm Event on Brays Bayou, Houston, Texas.

4. PARTICLE AGGREGATION AND THE FATE OF METALS

Once metals in runoff enter urban waterways they may undergo numerous transport and transformation processes. Dissolved metals may be sorbed to particle surfaces or taken up by biota. Metals that are complexed with natural organic matter may subsequently sorb to particle surfaces. Metals may also precipitate, dissolve, desorb, or participate in redox reactions. In most cases, the solid phases (sorbed or precipitated) are likely to play a critical role in determining the fate of metals that enter surface waters with urban runoff. In particular, the degree of aggregation that occurs in the runoff stream will directly affect the fate and transport of the particles, and thus the metals associated with them. Aggregation of sub-micron particles may enhance the flux of metals to sediments as metals on smaller, unsettleable colloids are incorporated into larger, faster settling aggregates.

It is likely that most of the suspended particle volume in urban waterways is present as aggregates rather than as individual particles, and that significant aggregation occurs during relatively short periods of time. Direct observation of suspended materials in rivers affected by urban runoff has shown that under ultrasound, the particle size distribution of this material is significantly shifted to the smaller sizes [22]. Also, measurements of particle aggregation rates in urban runoff waters under controlled laboratory conditions indicate that, at least in the case of the Houston watershed, particles exhibit a significant tendency to aggregate, as quantified by the mean particle size and the particle collision efficiency factor calculated from particle number distributions over time. If particle-particle collisions are predominantly shear-induced, and if the total particle volume remains constant, then the kinetics of initial particle aggregation can be approximated as first-order. In addition, if the velocity gradient in a mixed system is approximated by the Camp velocity gradient, G, then particle number concentration, n, over time, t, during the early stages of aggregation can be calculated as:

$$\frac{n}{n_o} = \exp(-\alpha \frac{4}{\pi} \phi G t) \qquad (1)$$

where α is the collision efficiency factor, ϕ is the particle volume fraction, and n_o is the initial particle number concentration. For known values of G, a plot of the natural log of particle concentration in an aggregating suspension versus time should fall on a straight line with a slope equal to the collision efficiency factor (Figure 5).

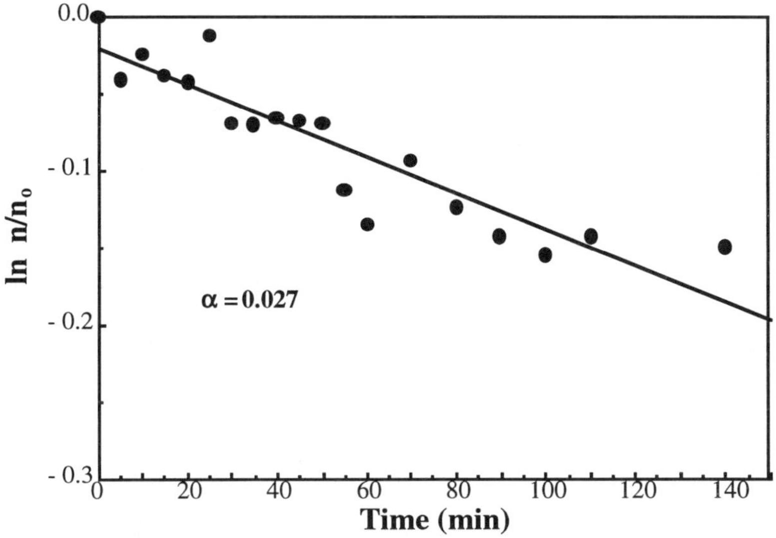

***Figure 5.** Log Particle Number Concentration Over Time in Stormwater Runoff Collected from Brays Bayou at Main Street, December 5, 1994. The runoff suspension was mixed at an intensity, G, of 70 per second.*

A value of α on the order of 10^{-2} indicates that these particles are reasonably stable; for every 100 particle collisions, only 2 or 3 result in an aggregation event. However, relatively high particle concentrations present in storm water runoff resulted in an approximately 15% reduction in particle number in a little over two hours of mixing, after which time large visible aggregates became apparent. Thus, significant aggregation may occur even in suspensions of relatively stable particles.

Particle aggregation can be represented in greater detail by considering Brownian aggregation and differential settling in addition to shear-induced particle collisions. A system of differential equations in which each equation describes changes in particle number within a given size class over time is solved simultaneously. Simulations of particle aggregation over time using measured particle size distributions as the initial conditions and assuming a value of 0.03 for the collision efficiency factor yield significant changes in the particle volume distribution after just an hour of aggregation (Figure 6). It appears likely that under the conditions typical of urban runoff events, colloids rapidly aggregate to sizes where breakup begins to control the size distribution. Thus, small colloidal materials and associated metals may be rapidly "pumped" into the larger particle size classes during runoff events. These larger materials will be deposited closer to the entrances of lakes and estuaries, a result that will be enhanced by reduced flow velocities and higher salinities.

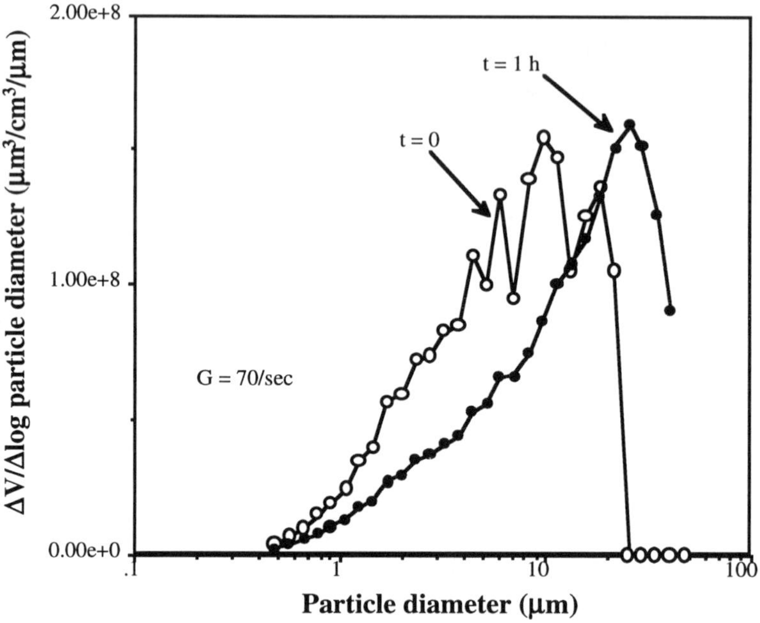

Figure 6. Simulated Change in Particle Distribution Over One Hour for a Suspension of Particles Entering an Urban Waterway During a Storm Event. A particle collision efficiency factor of 0.03 was assumed.

5. SUMMARY

Urban runoff is often a significant source of metals to surface waters. A single storm event may deliver the equivalent of several days of metals loading under conditions of background flow. During these storm events, the concentrations of different metals entering urban waterways may vary over time. Much of this variance appears to be explained by the relative tendency of metals to associate with the particulate phase or dissolved organic carbon. Most of the particle surface area present in urban runoff is in the colloidal size range. Once mobilized, particles in urban runoff are likely to quickly aggregate, altering the transport and fate of suspended solids and associated metals. Based on a limited survey of existing data, zinc concentrations in urban runoff would seem to be largely predictable based on per capita fuel consumption or urban density. The higher concentrations of zinc observed in metropolitan areas with higher *per capita* fuel consumption are consistent with the notion that tire wear releases zinc to the environment.

REFERENCES

1. Newell, C.J., H.S. Rifai and P.B. Bedient, "Characterization of Non-Point Sources and Loadings to Galveston Bay," Galveston Bay National Estuary Program (1991).
2. Sartor, J.D. and G.B. Boyd "Water Pollution Aspects of Street Surface Contaminants," USEPA (1972).
3. Sartor, J.D., G.B. Boyd and F.J. Agardy, "Water Pollution Aspects of Street Surface Contaminants." *J. Water Pollut. Control Fed.* 46:458-467 (1974).
4. Colston, N.V.J., "Characterization and Treatment of Urban Runoff," EPA-670/2-71-09, U.S. Environmental Protection Agency (Springfield, 1974).
5. Newton, C.D., W.W. Shephard and M.S. Coleman, "Street Runoff as a Source of Lead Pollution," *J. Water Pollut. Control Fed.* 46:999-1000 (1974).
6. Hedley, G. and J.C. Lockley, "Quality of Water Discharged from an Urban Motorway," *J. Water Pollut. Control Fed.* 74:659-674 (1975).
7. Hunter, J.V., J. Balmat, W. Wilber and T. Sabatino, "Hydrocarbons and Heavy Metals in Urban Runoff," in *Proc. Conf. Urbanization, Stormwater Runoff and the Aquatic Environment*, George Mason University (1981).
8. Wilber, W.G. and J.V. Hunter, "Contributions of Metals Resulting from Stormwater Runoff and Precipitation in Lodi, New Jersey," in *Urbanization and Water Quality Control*, American Water Resources Association (1975).
9. Whipple, W.J. and J.V. Hunter, "Nonpoint Sources and Planning for Water Pollution Control," *J. Water Pollut. Control Fed* 49:15-23 (1977).

10. Whipple, W.J., J.V. Hunter and S.L. Yu, "Effects of Storm Frequency on Pollution from Urban Runoff," *J. Water Pollut. Control Fed.* 49:2243-2248 (1977).

11. Whipple, W.J., J.V. Hunter and S.L. Yu, "Runoff Pollution from Multiple Family Housing." *Water Res. Bull.* 14:288-301 (1978).

12. Pardo, R., E. Barrado, L. Perez and V. Marisol, "Determination and Speciation of Heavy Metals in Sediments of the Pisuerga River," *Water Res.* 24:373-379 (1990).

13. Fuller, C.C., J.A. Davis, D.J. Cain, P.J. Lamothe, T.L. Fries, G. Fernandez, J. Vargas and M.M. Murillo, "Distribution and Transport of Sediment-bound Metal Contaminants in the Rio Grande de Tarcoles, Costa Rica (Central America)," *Water Res.* 24:805-812 (1990).

14. Calvo, C., R. Donazzolo, F. Guidi and A.A. Orio, "Heavy Metal Pollution Studies by Resuspension Experiments in Venice Lagoon," *Water Res.* 25:1295-1302 (1991).

15. Boughriet, A., B. Ouddane, J.C. Fischer, M. Wartel and G. Leman, "Variability of Dissolved Mn and Zn in the Seine Estuary and Chemical Speciation of These Metals in Suspended Matter," *Water Res.* 26:1359-1378 (1992).

16. Lo, C.K. and Y.S. Fung, "Heavy Metal Pollution Profiles of Dated Sediment Cores from Hebe Haven, Hong Kong," *Water Res.* 26:1605-1619 (1992).

17. Williamson, R.B., Urban Stormwater Quality I. Hillcrest, Hamilton, New Zealand," *New Zealand J. Mar. Fresh. Res.* 19:413-427 (1985).

18. Tanizaki, Y., T. Shimokawa and M. Yamazaki, "Physico-Chemical Speciation of Trace Elements in Urban Streams by Size Fractionation," *Water Res.* 26:55-63 (1992).

19. Wilber, W.G. and J.V. Hunter, "The Impact of Urbanization on the Distribution of Heavy Metals in Bottom Sediments of the Saddle River," *Water Res. Bull.* 15:790-800 (1979).

20. Wilber, W.G., J.V. Hunter and J. Balmat, "Zinc in Urban Storm- and Wastewaters," in *Zinc in the Environment, Part I: Ecological Cycling*, J.O. Nriagu, Ed., pp. 230-263. John Wiley & Sons (New York, NY 1980).

21. Characklis, G.W., "Characterization of Particles, Metals and Water Quality in Urban Runoff," Master's Thesis, Rice University, Houston (1994).

22. Droppo, I.G. and E.D. Ongley, "The State of Suspended Sediment in the Freshwater Fluvial Environment: A Method of Analysis." *Water Res.* 26:65-72 (1992).

INDUSTRIAL DISCHARGES OF METALS TO WATER

James W. Patterson,[1] *Roger A. Minear,*[2] *Eduardo Gasca*[1] *and Chriso Petropoulou*[1]
[1]Patterson Associates, Inc.
Chicago, IL 60603
[2]Institute for Environmental Studies
University of Illinois at Urbana-Champaign
Urbana, IL 61801

1. INTRODUCTION AND HISTORICAL PERSPECTIVE

1.1 Introduction

The discharge of heavy metals and associated process substances dates back many decades and has been a matter of some environmental concern nearly as long. Because of this concern, treatment of heavy-metal wastewaters has been practiced for more than fifty years. In early years, principal concerns related to the impacts on receiving waters of high strength and highly toxic solutions. Notable among these were the various plating wastewaters containing cyanide and hexavalent chromium. Non-plating metal surface treatment operations in many cases also involved hexavalent chromium solutions at high concentrations.

Historically, the specific concentrations and types of metals, metalloids, and associated process chemicals provoking concern have expanded dramatically. Also, as manufacturing processes have changed, the range of industrial sources and chemical complexity of associated process solutions has expanded. An overview of industrial sources and concentrations is presented to illustrate these changes and those associated with the evolution in metal wastewater treatment technology.

Pollution control technologies have evolved from simple, end–of–pipe technology to alternatives that involve in-plant source control, recycle and recovery, and pollution prevention initiatives. Alternative manufacturing processes or actual materials substitution approaches can provide opportunities for avoidance of intractable problems.

The evolution of treatment technologies associated with major environmental legislation includes the treatment technologies associated with "Categorical Effluent Limitation Guidelines Compliance" and treatment technologies associated with "Water Quality Based Effluent Limitations." Advanced treatment technologies thus evolving must be examined in terms of the tradeoffs in 1) costs, 2) uncertainties in technical application, 3) hazardous waste generation, 4) opportunities for water conservation and reuse, 5) opportunities for by–product recovery and reuse, and 6) incidental removals associated with processes directed at other treatment targets.

1.2 Historical Perspective

Metals are elements, which can not be destroyed by treatment technologies. Metals in wastewater normally represent raw materials, utilized by many different types of industry. Inherent, and often unavoidable inefficiencies in the industrial manufacturing processes result in the release of metals from these processes. Table 1 presents example generic industrial origins of metals in wastewaters.

Table 1. Generic origins of metal–bearing wastes [1].
Spent processing baths, as from electroplating or etching and cleaning
Rinsewaters from processing baths
Air pollution scrubber waters, as from smelting and refining, or combustion
Ore and coal mine drainage
Raw material, product, or production equipment cleaning
Sluice or quench water as from flyash removal, or casting or extrusion quenching

Many industrially developed countries have imposed regulatory constraints on the release of toxic metals and other toxics into the water environment. In order to comply with these constraints, industries have installed, and operate, often highly efficient pollution control facilities.

These systems are essentially only concentration processes, however, to partially cleanse the relatively dilute wastestreams of the metals and to concentrate these metals into sludges or brine streams. There are myriad technologies available to achieve these concentration functions, and the final selection is typically based simply upon technological and economic factors. The technologies all share one characteristic, however, of producing a concentrated and highly contaminated residue for which only one management option has traditionally existed. That option has been land disposal, usually landfills. Occasionally, other land disposal methods such as deep well injection or evaporative impoundments have been utilized. Irrespectively, land becomes the ultimate repository of the residuals. Within such sites, metals may escape to the groundwater, and may undergo chemical or biochemical transformations which accelerate that escape.

Table 2 summarizes the principal technologies for metals control. The technologies can be divided into two broad categories:

- Conventional Metals Treatment Technologies
- Emerging Metals Treatment and Recovery Technologies

The technologies as listed in Table 2, Category A, are ranked by approximate frequency of application in metals pollution control. The level of utilization of the technologies assigned to Category B range from infrequent to no current full-scale application, and are listed without implication of being ranked for either current utilization or future potential.

The conventional treatment technologies are normally utilized for relatively non-selective treatment, and always generate a residue (e.g., a sludge) requiring management. In no instance is recovery an objective in the application of these conventional treatment technologies. The first four technologies in Category B are more current and may be utilized to some extent for selected targeted applications. The last three technologies are more emerging technologies and have undergone rapid evolution within the past decade, and indeed, as discussed below, differential precipitation, was recently defined by the U.S. EPA as the best practicable control technology currently available (BPT) for wastewater treatment for the Centralized Waste Treatment Industry. These and more advanced technologies are discussed in further detail in Section 3.0.

Table 2. Principal technologies for metals control [1].

A. Conventional Treatment Technologies (Ranked)	
	Precipitation
	Oxidation/Precipitation
	Reduction/Precipitation
	Concentration/Precipitation
	Secondary Co–precipitation
B. Emerging Treatment and Recovery Technologies (Unranked)	
	Evaporative Recovery
	Ion Exchange
	Membrane Separation
	Reductive Electrolysis
	Differential Precipitation
	Extractive Metallurgy
	Selective Adsorption

Differential precipitation shows significant promise with respect to (a) achieving high levels of effluent quality, and/or (b) promoting opportunities for pollution prevention and resource recovery. This approach is considered later in this paper, via case study, in Section 3.4.

The potential for achieving high levels of pollution control has been recently exploited by the U.S. EPA in defining BPT for the Centralized Waste Treatment Industry [2]. This industry receives industrial wastes from off-site for treatment or recovery. The EPA has defined several segments of this category, including Subcategory A: Facilities which treat, or treat and recover metals from metal–bearing wastes received from off-site. There are 55 such facilities in the U.S. For BPT, Subcategory A, the EPA has established the following model treatment sequence:

1. Selective Metals Precipitation
2. Pressure Filtration
3. Secondary Precipitation
4. Solid–Liquid Separation
5. Tertiary Precipitation
6. Solid–Liquid Separation

Where the metals streams contain cyanide, alkaline chlorination would precede Step 1, above. Based upon this application of BPT, the EPA has proposed effluent limitations as presented in Table 3. For comparison, current discharge limits for the State of Illinois as based on the performance of conventional technologies are also presented.

For most of the metals regulated, these BPT limits are predicated upon the performance of differential precipitation and are far more restrictive than metal discharge limits promulgated on the basis of performance of conventional treatment technologies.

Table 3. Proposed EPA BPT effluent limitations [2] and Illinois discharge limits [3].

Metal	U.S. EPA Daily Maximum mg/L	U.S. EPA Monthly Average mg/L	Illinois Monthly Average mg/L
Aluminum	0.72	0.16	-
Antimony	0.14	0.031	-
Arsenic	0.076	0.017	0.25
Barium	0.14	0.032	2.0
Cadmium	0.73	0.16	0.15
Chromium	0.77	0.17	1.0
Cobalt	0.73	0.16	-
Copper	1.0	0.23	0.5
Hexavalent Chromium	0.14	0.077	0.1
Iron	2.4	0.54	2.0
Lead	0.37	0.082	0.2
Magnesium	9.9	2.2	-
Manganese	0.18	0.039	1.0
Mercury	0.013	0.0030	0.0005
Nickel	5.4	1.2	1.0
Silver	0.028	0.0063	0.1
Tin	0.20	0.044	-
Titanium	0.021	0.0047	-
Zinc	1.2	0.27	1.0

2. INDUSTRIAL SOURCES AND CONCENTRATIONS

This section discusses the industrial sources of heavy metals discharged to water. The metals considered include cadmium, chromium, copper, lead, mercury, nickel, silver and zinc. The primary industrial sources of metal discharges include the metal mining industry (SIC 10), chemical and allied products (SIC 28), leather and leather products (SIC 31), fabricated metal products except machinery and transportation equipment (SIC 34), electronic and electric equipment (SIC 36), transportation equipment (SIC 37) and electric, gas and sanitary (SIC 49). Other industries that are known to discharge metals during processing include the Plastic Materials and Synthetic Organic Chemicals (chromium and zinc), Petroleum Refining and Petrochemicals (chromium), Nonferrous Metals (cadmium, copper, lead and zinc) and Textiles (chromium) [4].

Table 4 presents a tabulation of metal discharges by industries with full-scale treatment systems, given a range of metal influent concentration from a low range of non-detectable (ND) to 0.1 mg/L to a high range of 100 to 1,000 mg/L. This table was prepared from information obtained from the U.S. Environmental Protection Agency's Risk Reduction Engineering Laboratory (RREL) Treatability Database, Version Number 5.0 [5].

For cadmium, influent concentrations below or at 0.1 mg/L are typical for the chemical and allied products, the leather and leather products, and electric, gas and sanitary industries.

Industrial wastewater discharge sources of cadmium that range from more than 0.1 to 1.0 mg/L include the chemical and allied products. Other industrial sources reported to discharge at this cadmium concentration range include nonferrous metals manufacture (copper, lead, zinc and copper and zinc smelting, lead and zinc refining), iron and steel manufacturing, paint and ink formulation and rubber processing [6].

Cadmium concentration discharges above 1 mg/L are typical of the fabricated metal products and electronic and electric equipment. Specific example processes include plating rinsewaters and rinse dragout, bright dip and passivation plating baths and metal finishing.

RREL treatability data for aqueous trivalent chromium are not available in the database file. Information for trivalent chromium was obtained from Patterson [6]. Leather and leather products, fabricated metal products and electronic and electric equipment are the major industrial sources of chromium.

Trivalent chromium in industrial effluents may also result from the reduction of chromium (VI) bearing wastewaters. This is specifically the case for plating and plating bath rinse wastewaters. Industrial sectors that are known to use trivalent chromium include glass, leather, ceramics, photography, inorganic pigments and production of chromium metal processes [6]. Wet scrubber wastewaters from various metal smelting processes are also known to contain trivalent chromium. Other trivalent chromium related discharges include cooling towers, tanneries, steel mill effluents, titanium dioxide and sodium dichromate production [6].

Industrial discharges of hexavalent chromium include the lumber and wood products, chemicals and allied products, stone, clay and glass products, and machinery (except electrical) [7].

Table 4. Industrial sources of metal discharges to water.[1]

Range of Influent Concentration (mg/L)	Metal Discharges by Standard Industrial Classification (SIC)							
	Cd	Cr[2]	Cu	Pb	Hg	Ni	Ag	Zn
ND – 0.1	28 31 49	36	28 31	28 31 36	28 31	28 31 36	28 34	28
>0.1 – 1.0	28	34	28 31 36 37 49	28 31 49	28	28 31 34 37 49	28	28 31 49
>1.0 – 10	34 36	31 34	10 28 34	38 31 34	--[3]	28 34	--	28 34
>10 – 100	34 36	31 34	28 34	--	--	28 34	U[4]	10 28 34 37 49
>100 – 1000	--	--	--	37	--	36 49	U	28

[1]Information extracted from Aqueous data file of the RREL Treatability Database, Ver. No. 5.0 [5].

[2]RREL Database for trivalent chromium is not reported. Influent chromium data presented in Table A includes both trivalent and hexavalent chromium. Treatability data for trivalent chromium were obtained from Patterson [6].

[3]No industrial sector reported for that range of metal influent concentrations.

[4]Unspecified SIC sectors.

Copper-bearing wastewaters are associated with the metal mining, chemical and allied products, leather and leather products, fabricated metal products, electronic and electric equipment, transportation equipment, and electric, gas and sanitary. Specific industrial sources known to discharge copper-bearing wastewaters include primary and secondary copper smelting industries, copper plating rinse, appliance manufacturing, circuit board and integrated circuit manufacture, metal finishing, silver and brass plating, anodizing processes, brass mill rinse and pickle, copper mills, copper ore extraction, paint formulating, wood preservation, copper sulfate manufacture, porcelain enameling and air pollution control wet scrubbing systems.

Lead is used in many applications. The major industrial application is the lead battery industry. Other applications include pigments, leaded glass, fuels, photographic materials and explosives. Lead bearing wastewaters reported in the RREL database are attributed to the chemical and allied products, leather and leather products, fabricated metal products, electronic and electric equipment and electric, gas and sanitary industrial sectors [5]. Specific processes associated with lead discharges include plating, printed circuit board, glass, porcelain enameling, chloro–alkali plants, ammunition plants, tetraethyl lead manufacture, spent ink, paint and pigment manufacture, textile dyeing, rubber hose manufacture and foundries [6].

Industrial discharges of mercury include wastewaters generated from the chloro–alkali industry, electronic and electrical, explosives manufacture, photographic industry and preservative industry [6]. Paper mills, fertilizer mills, and smelting plants are also known to discharge low concentrations of mercury (0 to 40 μg/L). Water–based paint manufacturing, fluorescent lamp production, textile mills, secondary lead battery recovery and rubber processing generate wastewaters with a mercury content at concentrations below 1 mg/L [6]. Mercury is known to be found in trace amounts in industrial grade caustic and sulfuric acid.

Nickel bearing wastewaters are generated by almost each industrial sector cited in this paper. Metal plating and metal finishing operations are probably the most common sources of these wastewaters. Other typically known nickel wastewaters dischargers include those from the mine drainage, coal storage pile drainage, fly ash ponds, boiler cleaning, dye house, paint and ink formulating, nickel sulfate, copper sulfate, titanium dioxide and sodium dichromate manufacture.

Industrial wastewater dischargers associated with silver include the porcelain industry, mirror industry, photographic processing and electroplating [6]. Silver discharges below 1 mg/L are typical of the paint manufacture, textile dyeing and mills, iron and steel, commercial laundries and steam electric power plants. Discharges above 1 mg/L are related to photographic chemicals and processing, and silver plating.

The most common zinc bearing wastewaters are from the iron and steel industry, galvanizing lines, zinc and brass metal works and plating, rayon yarn and fiber production. Other zinc wastewater streams are generated by the latex rubber products, vulcanized fiber, wet scrubbers from municipal refuse incinerator and scrap steel cupola, chemical and allied industry, paint, ink, and dye formulating, petroleum refining and nonferrous primary and secondary metal sectors (lead, copper and zinc).

3. END-OF-PIPE TREATMENT: TECHNOLOGIES AND ASSOCIATED EFFLUENT CONCENTRATIONS

3.1 Treatment Technologies for Categorical Effluent Limitations Guideline Compliance

In 1972 federal water pollution legislation was enacted under Public Law 92–500. Section 402 (the National Pollutant Discharge Elimination System) and Section 306 (related to the National Standards of Performance) are of relevance to industrial waste discharges. Other sections of this law direct the EPA to establish wastewater discharge guidelines. "Best practicable control technology" (BPT) was required for industrial use by July 1977. By mid 1983 "best control technology economically available" (BAT) was to be implemented. U.S. EPA established categorical effluent limitation guidelines for specific industrial sources (see example in Table 5 for a leather tanning sector). These guidelines are based on several factors and conditions such as whether the industrial source is a new or existing source, if the source discharges treated effluent directly (NPDES) to "waters of the United States" or indirectly (Pretreatment) through Publicly Owned Treatment Works (POTW). Categorical effluent limitation guidelines are normally stricter for new sources discharging directly to waters of the United States via a receiving stream that has a ten year minimum seven day average flow of zero (7Q10 flow = 0 gpd). In most cases, other discharge limits (in addition to categorical limits) are established due to the water quality of the receiving stream. In contrast, the least stringent discharge limitations are normally associated with indirect discharges from existing sources to POTW systems.

Patterson [6] reports that targeted wastewater treatment technologies for heavy metals are essentially those employed in the plating and metal processing industries. Common technologies include chemical precipitation as hydroxide, sulfide or carbonate, coagulation, co-precipitation and ion exchange. Other technologies used include electrolytic and evaporative recovery processes and adsorption.

Hexavalent chromium reduction followed by alkaline precipitation is a well documented [6] technology. Other hexavalent chromium treatment technologies include ion exchange, electrochemical reduction and evaporative recovery.

Hydroxide precipitation of a waste containing mercury is normally ineffective. Patterson [6] reports sulfide precipitation, ion exchange, adsorption and coagulation, and reduction of ionic mercury to elemental form with removal by filtration as common treatment technologies for mercury.

Table 6 presents a summary of wastewater treatment technologies and associated maximum treatment efficiency reported in the RREL database [5]. Information is organized by parameter and influent metal concentration range from low (0 to 0.1 mg/L) to high range (>10 – 100 mg/L). A brief inspection of the technologies reported shows that many of the technologies are not considered technologies specific for heavy metals treatment (e.g., chemical precipitation or coagulation) and that incidental treatment of metals is a common full-scale industrial phenomenon. The subject of incidental treatment is discussed later in this chapter. Information compiled in Table 6 is limited to the aqueous industrial wastewater matrix for full-scale treatment systems. For a given technology that in the RREL database presents several percent removal values, the maximum value reported is considered in this paper. The abbreviation of technologies used by the RREL database is also adopted in this paper and a summary table of these abbreviations is presented at the end of the chapter in Appendix A.

3.2 Treatment Technologies for Water Quality Based Effluent Limitations

Industrial discharges directed to a zero or low flow receiving stream are normally required to meet stringent wastewater discharge limits that are based, in most cases, on the receiving water quality and flow of the receiving stream (not on BPT or BAT technologies). Even though federal, state and local guidelines may allow a "mixing zone" approach to establishing the wastewater discharge permit limits, antidegradation policies may not allow enough latitude for existing industrial users to renegotiate permit discharge limits. This is especially the case when treated effluents have consistently discharged effluent quality that represents much lower concentrations of regulated parameters than those allowed in the discharge limits.

Table 7 presents a summary of the freshwater quality criteria for the heavy metals presented in Table 4. The concentrations given are normally used by permit preparers to arrive at alternative water quality based NPDES discharge limits. If one considers the use of an existing industrial source discharging to a zero flow 7Q10 stream, the allowable heavy metals discharge concentration is of the same order of magnitude as that presented in Table 7. Industrial dischargers under these circumstances turn to research – oriented initiatives to select wastewater treatment technologies that become almost site-specific applications and in some cases, industry is forced to renegotiate or resolve this type of condition via other measures such as discharging to a receiving stream that would allow for some "mixing zone" or directing discharges to nearby POTWs. A typical example of industrial discharge history was recently published [8]. Here, the authors describe how the original textile mill wastewater treatment plant (constructed in 1952) discharged into a tributary of the west branch of the Susquehanna River at Woolrich, PA. First, the company decided in 1974 to connect into the POTW, which later adopted a new pretreatment ordinance in 1991. This new ordinance caused the existing industrial pretreatment facility to not meet new discharge limits for chromium and zinc, and other classical pollutants. The adoption of the ordinance triggered negotiations between the industrial user and the POTW, and an evaluation of alternative innovative treatment technologies.

After evaluating alternatives, the company opted to replace the existing pretreatment facility with an electro-chemical treatment system. The company also developed short–term goals that included meeting compliance with discharge limits, and long–term goals which envisioned partial and complete recycle of treated wastewater.

3.3 Trade–offs in Applications of Advanced Technologies

As discussed in Section 1.2, there are a number of recognized and emerging advanced technologies for metals control. These include evaporative recovery, ion exchange, membrane separation, reductive electrolysis, differential precipitation, extractive metallurgy, and selective adsorption.

Table 5. Applicable categorical effluent limitations summary for the hair save, chrometan, retan–wet finishing category of the leather and leather tanning industries.

	NPDES (Direct Discharge)				Pretreatment	
	BAT (Existing)		New Source		Existing and New Sources	
	(lb/1,000 lb of raw material)				mg/L	
	Maximum for Any One Day	Maximum for Monthly Average	Maximum for Any One Day	Maximum for Monthly Average	Maximum for Any One Day	Maximum for Monthly Average
Total Chromium	0.21	0.08	0.16	0.06	12	8
Sulfide	NR[1]	NR	NR	NR	24	NR
BOD 5	NR	NR	6.0	2.7	NR	NR
TSS	NR	NR	8.7	4.0	NR	NR
Oil & Grease	NR	NR	2.5	1.1	NR	NR
pH	(2)	(2)	(2)	(2)	(3)	(3)

[1]NR – Not regulated under categorical standards.
[2]Within the range of 6.0 to 9.0.
[3]Within the range of 7.0 to 10.0.

Table 6. Treatment technologies and associated removal for metals discharged from industrial sources.

Influent Concentration Range mg/L	Technology (Defined in Appendix A)	Maximum % Removal Reported
	Cadmium	
ND – 0.1	CAC	> 80
	ChOx (Cl)	> 25
	RA (Batch) + Fil	> 38
	SS	> 56
>0.1 – 1.0	AL	> 75
	AS	77
	ChPt (Batch) + Fil	99
>1.0 – 10	CAC w Fil	95
	ChPt	96
>10 – 100	ChOx/Pt (Batch)	72
	ChPt	86
	Chromium (VI)	
ND – 0.1	CAC W Fil	75
>0.1 – 1.0	ChPt + Fil	> 90
>1.0 – 10	ChRed	> 93
>10 – 100	ChPt + Fil	> 99
	Copper	
ND – 0.1	AL + Fil	> 76
	AS	58
	CAC	94
	ChOx (Cl)	14
	GAC	80
	PACT	> 76
	SExt	47
	SS	22
>0.1 – 1.0	AL	83
	AS	65
	CAC	88
	CAC w Fil	70
	ChPt + Fil	98
	Fil	26
	GAC	80
	RA	82
	RA (Batch) + Fil	95
	SS + GAC	78
	Sed	> 83
	Sed + AS	81

Table 6. Treatment technologies and associated removal for metals discharged from industrial sources (continued).

Influent Concentration Range mg/L	Technology (Defined in Appendix A)	Maximum % Removal Reported
	Copper (continued)	
>1 – 10	AS	98
	ChPt	> 99
	ChPt w Fil	90
>10 – 100	ChOx/Pt (Batch)	68
	ChPt	96
	ChPt (Batch) + Fil	99
	Lead	
ND – 0.1	AS	> 24
	CAC	50
	CAC w Fil	43
	PACT	> 13
>0.1 – 1	AS	> 99
	CAC	99
	ChOx (Cl)	49
	ChPt + Fil	93
	PACT	17
	RA (Batch) + Fil	> 57
>1 – 10	AL	> 68
	AS	> 97
	ChPt	> 82
	ChPt (Batch) + Fil	98
>100 – 1,000	ChPt + Fil	> 99
	Mercury	
ND – 0.1	AL	> 20
	AS	> 50
	CAC (Batch)	87
	ChOx (Cl)	3
	GAC	> 23
	RA (Batch) + Fil	> 85
	SS + GAC	85
	Sed	> 9
>0.1 – 1.0	ChPt (Batch) + Fil	89
	Nickel	
ND – 0.1	AS	> 59
	CAC	75
	CAC w Fil	20
	GAC	33

Table 6. Treatment technologies and associated removal for metals discharged from industrial sources (continued).		
Influent Concentration Range mg/L	Technology (Defined in Appendix A)	Maximum % Removal Reported
	Nickel (continued)	
>0.1 – 1.0	AL	> 76
	CAC	32
	ChOx (Cl)	19
	ChPt + Fil	96
	Fil	44
	GAC	> 73
	SExt	90
	SS	11
	Sed	> 67
>1.0 – 10	ChPt	98
	GAC	24
>10 – 100	ChOx/Pt (Batch)	81
	ChPt	88
	ChPt (Batch) + Fil	> 98
>100 – 1,000	ChPt + Fil	> 99
	RO	97
	Silver	
ND – 0.1	AS	> 79
	ChPt	> 80
>0.1 – 1.0	AL	> 65
>10 100	ChPt w Fil	> 99
>100 – 1,000	ChPt w Fil	> 99
	Zinc	
ND – 0.1	AS	22
	GAC	> 86
>0.1 – 1.0	AL	60
	AL + AS	81
	AL + Fil	18
	AS	98.5
	CAC	89
	Fil	80
	GAC	74
	PACT	82
	SS	15
	Sed	49
	Sed + AS	69

Table 6. Treatment technologies and associated removal for metals discharged from industrial sources (continued).

Influent Concentration Range mg/L	Technology (Defined in Appendix A)	Maximum % Removal Reported
Zinc (continued)		
>1 – 10	CAC	69
	ChPt	93.7
	GAC	> 95.1
	SS + GAC	13
>10 – 100	AS	57
	ChOx/Pt (Batch)	98.2
	ChPt	> 99
	ChPt + Fil	> 99
>100 – 1,000	CAC	> 99

Evaporative recovery involves the concentration of a wastestream by evaporation of water. Operating costs are high, and closely tied to the volume of water that must be extracted to produce a brine of acceptable concentration. Due to high energy requirements, energy efficient multiple–effect evaporators have become commercially available. While the multiple–effect evaporators are more energy efficient than single–effect evaporators, they also represent a significant increase in capital cost above the single–effect models.

The principal concept underlying evaporative recovery is that the concentrate obtained can be returned directly to the manufacturing process. However the concentrate will contain, and through reuse gradually increase in, secondary non–volatile contaminants. In a total closed loop recovery process, these contaminants can build up to a level which would deteriorate product quality. In one example application, for chromic acid rinsewater, iron and trivalent chromium must be removed by cation exchange prior to evaporative recovery of the chromate solution [10].

Ion exchange is a well established technology. Its potential for metals pollution control is well recognized, but its application is limited. Basically for pollution control, ion exchange serves one of two functions. The first is tertiary or polishing treatment of an effluent before discharge. The ion exchange resin cannot be heavily loaded if ion leakage into the effluent is to be prevented, and the consequent resin regenerant will not be sufficiently concentrated in metal to justify recovery. Recovery is further hampered by the relatively non–selective nature of most commercially available resins, resulting in the accumulation in the regenerant of secondary and undesired contaminating ions. For some wastewaters, irreversible resin fouling and associated lost of exchange capacity is also a problem.

Table 7. National Freswater Quality Criteria for specific heavy metals[1] (µg/L).

	H[2]	Cd	Cr(III)	Cr(VI)	Cu	Pb	Hg	Ni	Ag	Zn
Four–Day Average	50	0.66	120	11[3]	6.5	1.3	0.012[3]	56[4]	–	47[4]
	100	1.1	210	–	12	3.2	–	96[4]	–	
	200	2.0	370	–	21	7.7	–	160[4]	–	
One–Hour Average	50	1.8	980	16[3]	9.2	34	2.4[3]	1,100[5]	1.2[5]	180[5]
	100	3.9	1,700	–	18	82	–	1,800[5]	4.2[5]	320[5]
	200	8.6	3,100	–	34	200	–	3,100[5]	13[5]	570[5]

[1]Information compiled from Quality Criteria for Water [9].

[2]H–Hardness (mg/L as $CaCO_3$).

[3]Concentration value independent of hardness.

[4]24–hour average.

[5]At any time.

The second type of application of ion exchange is for direct recovery. Examples include recovery of silver, and recovery of chromate in plating wastewaters. This latter application involves a complicated sequence of three ion exchange columns in series. However, if the chromic acid produced in the resin regeneration step is too concentrated, resin matrix oxidative damage occurs. Thus the concentration of chromic acid recovered through regeneration is governed in part by the chemical instability of the matrix of the resin used.

Membrane separation has been widely considered for pollution control and resource recovery. Excluding ultrafiltration, which is more precisely a physical separation process, the principal commercially available membrane processes are electrodialysis and reverse osmosis. The perceived advantages and established disadvantages of both processes have been widely reported. Disadvantages primarily involve high capital and/or operating costs, non–selective nature of the separation process, membrane fouling, biological, chemical or physical deterioration of the membrane, and inability to economically achieve sufficient brine concentration or purity to justify recovery.

Reductive electrolysis is well established for certain metals recovery applications. The technology is varied, ranging from application of imposed electrical potential to plate out and recover cadmium from plating rinsewaters or copper from spent printed circuit board baths (electrowinning), to spontaneous electrolysis employing a sacrificial anode such as scrap iron. This latter application is frequently termed cementation, and has its widest application for copper recovery. Except for application for recovery of precious metals, or for oxidative destruction of spent cyanide baths, imposed reductive electrolysis is primarily applied for copper (and occasionally cadmium) recovery. Process inefficiencies and associated economics limit application to relatively concentrated solutions.

Extractive metallurgy, in various modifications, appears promising for recovery of metals from wastewater treatment sludges [11]. This potential will be enhanced by application to relatively clean sludges, in contrast to the complex sludges typically produced by end–of–pipe treatment of combined wastewater. Differential precipitation can enhance the feasibility of extractive metallurgy for metals recovery. Currently there are three approaches to the application of extractive metallurgy. Pyrometallurgy involves the application of heat, and is an ancient science first used more than six thousand years ago to initiate the Bronze Age. Its principal applications today, of course, are to smelt ores to produce virgin metals. Ironically, many wastewater treatment plant sludges are richer in target metals than are the ores from which the metals were originally extracted. However, the wastewater treatment plant sludge matrices are often not acceptable for direct pyrometallurgical extraction, in existing smelters. Current applications of pyrometallurgy for metals recovery are primarily restricted to precious metals, and some copper residues. The second approach to extractive metallurgy is hydrometallurgy. The concept involves dissolution (leaching) of the metal from the solid matrix, followed by a recovery stage which may involve leachate precipitation, crystallization, or reductive electrolysis. The term hydrometallurgy is somewhat of a misnomer. Although water may be involved in the dissolution step, many alternative leachants (e.g., acids) are also used. More precisely, then, extraction by a leachant must be considered to be the initial step in hydrometallurgy. Currently the major focus of hydrometallurgy is on metals extraction from virgin ores, and a variety of processes have been demonstrated. The third extractive metallurgy approach is biohydrometallurgy. This process involves the application of acclimated microorganisms (mostly bacteria) to enhance selective metal (s) leaching from mining areas. After leaching, common metallurgy concentration processes are followed. Concentration after the leaching step is often necessary, to achieve cost effective metals recovery. Concentration techniques can include liquid membrane technology, ion exchange or other technologies.

Selective adsorption, utilizing biological materials, mineral oxides, or polymer resins, has generated much excitement among researchers and practitioners of environmental engineering and science [11]. An alternative to selective adsorption is broad stroke adsorption, followed by selective (e.g., chromatographic) elution. Both approaches are aimed at separating a complex wastestream into a better defined set of components. Selective adsorption must be differentiated from a polishing technology such as ion exchange or co-precipitation. These latter approaches are normally targeted toward effluent polishing, as opposed to metals recovery.

Differential precipitation involves multi–step titration of a complex wastewater, to form and precipitate specific metal salts at selected titration points. Conceptually, the process could be applied to purge a wastewater of undesired species by differential precipitation, perhaps followed by application of a recovery process for the metal(s) remaining in solution. By example, this strategy is used at one plant to strip copper from a mixed wastewater, ahead of a precious metal recovery step [12]. The differential precipitation of copper was performed at pH 5, achieved through addition of caustic. Alternatively, differential precipitation could be employed to precipitate a metal targeted for recovery. The metal which precipitates could then be extracted for recovery. As discussed earlier, the U.S. EPA has designated differential precipitation as BPT to achieve an exceptional effluent quality.

Typically, thermodynamic modeling is applied for more complex wastewaters, in an attempt to predict key points in the differential precipitation titration process. Experience to date has indicated that such modeling can not adequately account for process kinetics or species interaction in real wastewaters, and experimental results usually fail to conform to model predictions [13,14]. However as discussed in the case study below (Section 3.4), selective precipitation when properly applied can yield valuable benefits in metals control technology.

Targeted differential precipitation can involve alternative objectives:

- forced formation of a desired salt (e.g., sulfide or carbonate precipitation),
- sequential stepwise precipitation for generation of relatively pure solid phases,
- sequential stepwise precipitation to achieve a very high effluent quality,
- formation of solid phases having desirable characteristics relating, for example, to particle size, density, or surface charge.

Examples of the first type, forced formation of the desired salt are numerous (Table 8). The strategy normally is applied in order to reduce target pollutant solubility in the wastewater, and thereby achieve desired effluent quality. When multiple metals are present, the resultant sludges may take on the characteristics of the least manageable components and, from the standpoint of sludge handling and metals recovery, no advantages are achieved [15,16]. Further, the forced salt formation processes are particularly sensitive to reaction stoichiometry, and secondary ion scavenging into the precipitate can interfere with process efficiency.

The second strategy, sequential stepwise precipitation, is quite promising. The technique involves the use of kinetic or other solubility controls to sequentially form and separate relatively pure salts or well defined salt mixtures. The technique has been utilized to sequentially precipitate and recover copper, and then silver, from a silver plating rinsewater [11]. Most initial attempts at sequential stepwise precipitation begin with equilibrium modeling. Unfortunately, kinetic features of selective precipitation frequently control real time results.

Table 8. Examples of selective salt precipitation [11].

Pollutant	Salt
Arsenate	Ferric Arsenate
Chromate	Barium Chromate
Cadmium, Lead	Metal Carbonate
Copper	Copper Oxide Copper Hydroxynitrate
Chromium, Copper, Lead Zinc	Metal Thioacetamide Metal Dibromo–oxime
Many Metals	Metal Sulfide

The third strategy is represented by the BPT limits proposed by the U.S. EPA for the Centralized Waste Treatment Industry (see Table 3) [2].

The fourth strategy seeks to control the discrete physical and/or adsorptive characteristics of the solid phase formed. Freshly precipitated salts typically have an electrical surface charge which results in adsorption of counter ions from the wastewater. The solid phase is thereby made less chemically pure, and less attractive for recovery. Surface charge management can suppress (or enhance) this adsorptive co–precipitation. In a like manner, this strategy may seek to achieve larger, or perhaps more dense (crystalline) solids, even if achieved within a matrix of rather gelatinous precipitates. Physical separation techniques such as differential sedimentation or centrifugation can then separate the intermingled solid phases.

3.4 A Case Study

This case study is for a large U.S. facility involved in recycling of copper scrap. Refinery and mill grade scrap is processed to yield high grade electrolytic copper or copper billets. Most of the refined copper is used on site for the manufacture of copper tubing, with the remainder of the refined product sold to other manufacturers. The industrial manufacturing activities comprise three elements:

1. Fire (pyrometallurgical) and electrolytic (hydrometallurgical leaching plus reductive electrolysis) refining of copper from purchased scrap,

2. Fire refining of scrap and casting of copper billets,

3. Manufacture of seamless copper tubular products, by extrusion.

Refinery grade copper scrap is melted and fire refined in furnaces, then cast into anodes which are the feed stock for the electrolytic refinery, which yields high grade electrolytic copper cathode. This product is used in subsequent billet casting operations, or sold to other users of high grade copper.

Electrolytically refined copper, together with internally generated scrap, is melted in a furnace and cast into copper billets. The copper billets are extruded into tubular products and hollow shells, which are then cold drawn into a variety of sizes serving the plumbing, refrigeration, air conditioning and automotive industries. This facility is the largest of its kind in the United States.

Many wastewaters are associated with the manufacturing processes. A primary type of waste stream is acidic wastewater containing metals. These wastewaters result from flue gas scrubbing operations at the smelting furnaces, from electrolytic refining including rinses, and from pickling operations. The quantities of fresh water traditionally used in this processes were significant, and could result in a significant amount of wastewater. Historically, the major wastewater constituent was copper. Copper was lost primarily as a result of the electrolytic refining operation. Significant copper was also lost via blowdown from the air pollution control systems on the smelting furnaces. Quantitatively, the raw electrolytic refining liquid wastes contained about 250,000 lbs per year of copper, the anode furnace scrubber blowdown contained approximately 17,000 lbs per year of copper, the wastewater from billet melting and casting contained about 15,000 lbs per year of copper, and the extrusion and piercing operations contained about 250 lbs per year of copper. In total, this represented about 150 tons per year of copper discharged in the effluent.

The primary constituent in the electrolytic refining process wastewater, other than copper, was nickel. Historically, approximately 15,000 lbs per year of nickel were lost via the wastewater. Other metals lost via the wastewater from this facility included zinc, iron, lead, and tin, with traces of numerous other metals. Approximately 115,000 gallons of concentrated sulfuric acid per year were used as makeup to electrolytic refining. Virtually all of this acid was lost to the wastewater.

The pollution control program involved three phases. The first phase addressed water conservation, and the second recirculation of partially treated water for reuse. The third phase involved targeted selective precipitation. The targeted selective precipitation was designed to accomplish three objectives. The first objective was to achieve a partially treated wastewater of quality acceptable for reuse or sequential use within the manufacturing facility. The second objective was to achieve a final effluent quality necessary to meet U.S. EPA categorical pretreatment standards for existing sources. The third and most innovative objective was to produce selective precipitation treatment sludges of sufficiently high quality for complete recovery, without need to landfill or otherwise dispose of any treatment residues (other than the treated effluent). The concept development program involved wastewater characterization studies, thermodynamic modeling to predict the effect of various parameters on the precipitation processes, and comprehensive benchscale studies to confirm or extend the results of the thermodynamic modeling. As a consequence of these studies and implementation of the recommended technology, this industrial facility was able to achieve the following objectives:

- Reduction in volume of regulated wastewater discharged by 78%,
- Complete recovery of process acids, representing over 1.7 million pounds per year,
- Reduction in discharge of regulated wastewater pollutants of over 99%,
- Total (100%) avoidance of wastewater treatment sludges disposal, and recycling of over 325,000 lbs per year of metals.

This case study, for this complex industrial facility, indicates that innovative approaches are technically and economically feasible.

4. ALTERNATIVE APPROACHES

4.1 In-plant Source Control and Pollution Prevention

Pollution prevention and waste minimization are terms that have been receiving increasing attention over the past decade. Many of the processes and approaches that are mentioned under this rubric in recent literature [17,18] have actually been around for several decades and were described in an assessment of treatment technologies in the early 1970s [19]. A number of these approaches have already been discussed above under Section 3.3 in the context of advanced technologies, although they also are currently categorized as waste minimization and pollution prevention methodologies. While not necessarily new in concept, the combination of improvements in technology, tighter regulations including on sludge disposal, and associated economic driving forces have made many of the reduction, recovery and recycling operations feasible in the 1990s, whereas in the 1970s, many of these considerations were cost effective only under selective applications.

In some cases, simple housekeeping reduces problems. The case study described above, also represents application of some of these principles of minimization and recovery. Collectively, these approaches represent material handling and simple operational changes, and recovery and reuse technologies. The practice of pollution prevention assessment is a formalized approach being advocated in many industries. An outline of this approach is presented in Table 9. Some aspects of this process for the plating industry relating to rinse waters, a major waste stream, are listed in Tables 10 and 11. These examples come from the Illinois Hazardous Waste Research and Information Center (IHWRIC), which has developed a major thrust in the pollution prevention arena.

Table 9. Elements of pollution prevention assessment.
A. Collect accurate and current waste generation information
1. Determine waste management costs
2. Determine raw material costs
3. Construct a preliminary waste flow diagram (mass balance) showing the location, volume and concentration of each waste stream
4. Determine data needs
B. Conduct a detailed plant survey of each area of the plant to determine exactly where in your processes wastes are generated and why
1. Correct/complete waste flow diagram
2. Use simple logs of material use/waste generation
C. Identify cost–effective source reduction techniques

4.2 Alternative Manufacturing Processes and Materials Substitution

Another approach being promoted focuses on changes in the manufacturing process itself. This can involve simple modifications that reduce the quantities of waste solutions as indicated in general above. It can involve changes in the composition of process chemicals such that the toxicity of attendant waste solutions is reduced, the treatment costs are reduced, recovery and recycle feasibility are enhanced or some combination of all these.

Table 10. Information needed on rinsewater generation [20].

Process Specific:
• Dragout rate from each process tank
• Composition of each process solution
• Actual operation parameters for each process tank
• Wastewater flow rate from each rinse tank
• Use of flow control techniques (conductivity cells)
• Use of dragout reduction techniques
• Actual operational procedures
• Location of each process tank
• Dump schedule for each process tank
• Manual or automatic line/barrel or rack plating
• Intermittent waste streams (cleaning, purification, dumps, etc.)
• Production rates (such as square feet processed per hour)
• Rack/barrel design and condition

Altmeyer [17] has examined the issue with respect to substitutions in the metal coating industry. Table 12 lists seven recommended changes discussed in his paper. Each of these changes offers advantages in toxicity reduction or removal, treatment cost reduction by elimination of specific treatment requirements, removal of reporting requirements, reduction in corrosion generated secondary waste constituents, to name a few. This is not to say such changes are without compromise in product quality or process reliability.

The above examples relate specifically to the plating industry. Other surface treatments produce significant metal waste problems. Cushnie [18] in addition to covering much of the same ground as Atlmeyer on plating operations, also discusses approaches to reducing or eliminating the use of hexavalent chromium in other metal finishing operations, such as anodizing, bright dipping, deoxidizing/desmutting, and stripping. Some cases involve use of trivalent chromium and others eliminate chrome usage altogether. Again, not all changes were universally successful. Other surface preparation or conversion processes which do not involve chrome may contribute other metal wastewaters directly or as a result of their action on the treated metal surface. These processes are also targeted for modification and discussed in Cushnie [18].

Table 11. Dragout reduction and recovery techniques [20].
• Increase part drain time
• Use low pressure air to blow solution off parts (air knife)
• Use fog or spray rinsing over the process tank
• Minimize concentrations of metal in plating bath
• Rack parts to maximize drainage
• Design racks to minimize drainage
• Maintain racks/barrels and coating
• Use drip bars to hold racked parts over plating tank
• Use drain boards between tanks
• Maximize hole size on barrels
• Rotate barrels above tank
• Place non-flowing rinse tank(s) after process tank

Table 12. Metal coating process bath formulation and procedure substitutions (from Altmeyer [17]).
1. Replace zinc and copper cyanide with non-cyanide forms
2. Replace hexavalent chrome with low concentration trivalent chrome
3. Substitute high pH nickel for copper strike on zincated surfaces and zinc die castings (eliminates cyanide)
4. Substitute acid salts for hydrochloric acid as a pre-dip prior to plating
5. Replace cadmium plating with zinc or zinc alloy processes to achieve enhanced corrosion resistance (also eliminates cyanide)
6. Replace hard chrome with nickel-tungsten-silicon carbide
7. Substitute carbon for electroless copper (circuit boards) (eliminates the use of formaldehyde)

Metals are also involved in other manufacturing processes as catalysts or components in product manufacturing. Many organic synthesis processes utilize metals as strong oxidizing or reducing agents. Among the metals used in this capacity are zinc, chromium, cerium, manganese, and osmium. Many organometallic reagents are also used in synthetic processes and many other organic and inorganic chemical reactions of commercial importance involve use of catalysts that are not always completely recovered or periodically become poisoned and require disposal. Attention in recent years has turned to development of new synthetic pathways where nonmetal catalysts, nonmetal redox processes can be used. In a recent ACS symposium, Epling and Wang [21] presented a method for sensitized photo oxidation processes to eliminate use of heavy metal oxide oxidants. Such approaches may become dominant processes of the future.

5. SUMMARY AND CONCLUSIONS

While industrial sources of metals have been reduced, it is clear that industrial practices still result in the discharge of metals to the environment. These discharges and past accumulations in various environmental compartments are the subject of most of this book. Clearly, as standards have become more stringent, old treatment technologies have improved, new technologies have become economically feasible and have led to lower effluent levels. We have tried to provide some insight into the evolution of the regulatory driving forces and the technologies resulting therefrom and the advantages and disadvantages of some of these technologies. That careful consideration and proper approaches can work quite effectively, technologically and economically, has been demonstrated in a case study. One can conclude that changes from the end-of-pipe treatment approach is feasible and that recycle and recovery is a starting point. However, active research into alternative approaches and waste reduction indicates that the future may ultimately reside in fundamental changes in manufacturing approaches and products produced.

APPENDIX A
TREATMENT TECHNOLOGIES TABLE
AQUEOUS DATA FILE

Treatment Technologies (Those with Data)	
AAS	Activated Alumina Sorption
AFF	Aerobic Fixed Film
AL	Aerobic Lagoons
API	API Oil/Water Separator
AS	Activated Sludge
AirS	Air Stripping
AlkHyd	Alkaline Hydrolysis
AlgIE	Algal Ion Exchange
AnFF	Anaerobic Fixed Film
BGAC	Biological Granular Activated Carbon
CAC	Chemically Assisted Clarification
ChOx	Chemical Oxidation (Parentheses show oxidation chemical; i.e., ChOx (UV) is ultraviolet light, ChOx (Cl) is chlorine, ChOx (Oz) is ozone, ChOx (H_2O_2) is peroxide, ChOx(ClO_2) is chlorine dioxide, and ChOx (Sur) is surfactant)
ChOx/Pt	Chemical Oxidation/Precipitation
ChPt	Chemical Precipitation
ChRed	Chemical Reduction
DAF	Dissolved Air Flotation
ED	Electrodialysis
Fil	Filtration
GAC	Activated Carbon (Granular)
IE	Ion Exchange (Parentheses show resin type; i.e. (A) is anionic, (C) is cationic, and (M) is mixed)
KPEG	Dechlorination of Toxics using an Alkoxide (Formed by the reaction of potassium hydroxide with polyethylene glycol (PEG400)
PAC	Powdered Activated Carbon
PACT	Powdered Activated Carbon Addition to Activated Sludge
RA	Resin Adsorption
RBC	Rotating Biological Contactor
RO	Reverse Osmosis
SBR	Sequential Batch Reactor
SCOx	Super Critical Oxidation
Sed	Sedimentation
SExt	Solvent Extraction
Soft	Water Softening
SS	Steam Stripping
TF	Trickling Filter
UF	Ultrafiltration
WOx	Wet Air Oxidation

REFERENCES

1. Patterson, J.W., "Metal Separation and Recovery," in *Metal Speciation Separation and Recovery Vol. I*, J.W. Patterson and R. Passino, Eds., pp. 63-93. Lewis Publishers (Chelsea, MI 1987).

2. U.S. EPA, "Development Document for Proposed Effluent Limitations Guidelines and Standards for the Centralized Waste Treatment Industry," EPA 821/R-95-006, U.S. Environmental Protection Agency, Office of Water (Washington, DC 1995).

3. IEPA, "Title 35: Environmental Protection Subtitle C: Water Pollution, Chapter I: Pollution Control Board," State of Illinois Rules and Regulations, IEPA, April (1988).

4. Nemerow, N.L., *Industrial Water Pollution Origins, Characteristics and Treatment,* Addison-Wesley Publishing Company (Reading, PA 1978).

5. U.S. EPA, RREL Treatability Database Version 5.0., Cincinnati, OH. (1993).

6. Patterson J.W., *Industrial Wastewater Treatment Technology Second Edition*, Butterworth Publishers (Stoneham, MA 1985).

7. Meltzer, M., M. Callahan, Y. Jensen, California Department of Health Services and U.S. Environmental Protection Agency, *Metal–Bearing Waste Streams Minimizing, Recycling and Treatment,* Noyes Data Corporation, Pollution Technology Review No. 196 (Park Ridge, NJ 1990).

8. Yohe, T.M. and J.E. Rich, "Textile Mill Builds for Today, Plans for the Future," *Industrial Wastewater*, May/June, (1995).

9. U.S. EPA "Quality Criteria for Water," EPA 440/5-86/001, U.S. Environmental Protection Agency, Office of Water, Regulations and Standards Division (Washington, DC 1986).

10. Yacumb, J. and K. Cachey, "Zero Discharge Chrome Recovery System Saves Energy and Recovers Valuable Resources," paper presented at the 56th Annual Conference, Water Pollution Control Federation (1983).

11. Patterson, J.W., "Metals Control Technology; Past, Present, and Future," in *Metal Speciation Separation and Recovery Vol. 2* J.W. Patterson and R. Passino, Eds., pp. 27-42. Lewis Publishers (Chelsea, MI 1990).

12. Gould, J., "Casebook," *Pollution Engineering*, July:10-11 (1984).

13. Jenke, D.R. and F.E. Dieble, "Recovery of Valuable Metals from Acid Mine Drainage by Selective Titration," *Water Res.* 17:1585-1590 (1983).

14. Karra, S.V., C.N. Haas, V. Tarre and H.E. Allen, "Kinetic Limitations of Selective Precipitation Treatment of Electronic Wastes," *Water Air Soil Pollut.* 24:253-265 (1983).

15. Horn, G., "The Effect of Carbonate on the Solubility of Cadmium and Copper," M.S. Thesis, Illinois Institute of Technology, Chicago (1983).

16. Jenista, J., "The Effect of Carbonate on the Solubility of Zinc and Copper," M.S. Thesis, Illinois Institute of Technology (Chicago, IL 1984).

17. Altmeyer, F., "A Comparison of Several Substitute Coatings as a Means of Pollution Prevention," presented at Chicago Metal Finishers Institute seminar, Chicago, IL, November (1992).

18. Cushnie, G.C., Jr., *Pollution Prevention and Control Technology for Plating Operations,* National Center for Manufacturing Sciences (Ann Arbor, MI 1994).

19. Patterson, J.W. and R.A. Minear, *Wastewater Treatment Technology*, IIEQ 71-4, State of Illinois (1971).

20. Hunt, G.E., "Hazardous Waste Minimization: Part IV Waste Reduction in the Metal Finishing Industry," *J. Air Pollut. Control Assoc.,* 38:672-680 (1988).

21. Epling, G.A. and Q. Wang, "Preparative Reactions Using Visible Light - High Yields from Pseudoelectrochemical Transformation," presented at 206th ACS National Meeting, ACS Division of Environmental Chemistry 33:328-329 (Chicago, IL 1993).

WASTE LOAD ALLOCATIONS FOR METALS

Lial Tischler
Tischler/Kocurek
Round Rock, TX 78664

1. INTRODUCTION

All states have either adopted numerical water quality criteria for potentially-toxic metals, or are subject to the U.S. Environmental Protection Agency's National Toxics Criteria (40 CFR 131) for metals, and are implementing these criteria in NPDES permits. Currently, most water quality-based permit limits for metals are based on simplifying assumptions including: simple mixing zone analyses without consideration of upstream and downstream sources; criteria that are assumed to be based on total metals or total recoverable metals rather than on dissolved metals; background concentrations of metals assumed to be zero; and no interaction between the metals in the water column and metals in the sediments. Some of these assumptions are conservative, while others are not. For receiving water bodies with relatively small numbers of significant point sources of metals and without significant nonpoint source metals loadings, the current methods of calculating water quality-based limits for metals are generally protective.

For surface water bodies with large numbers of point source discharges and/or significant nonpoint source loadings, more sophisticated waste load allocations for metals will be required to assure that permit limits are protective but not overly conservative. Waste load allocations for toxics, including metals, are designed to estimate the total maximum daily load (TMDL) for a receiving water segment. The TMDL is the maximum allowable loading of a pollutant to specified water body, from all sources (background, point sources, nonpoint sources) that will assure that the water quality standard for that pollutant is achieved at all locations under critical hydrologic conditions. This paper explores approaches to waste load allocations for metals, discusses the principal considerations that should go into selection of a particular approach, and cites several examples of waste load allocations that range from simplified to complex.

2. WASTE LOAD ALLOCATIONS

The objective of a waste load allocation (WLA) is to allocate available stream assimilation capacity among point and nonpoint sources to assure that the water quality standards are met under the full range of hydrologic and climatic conditions, with the exception of stream flows below the regulatory-defined critical low flow. Waste load allocations have been performed by state water quality agencies for many years to regulate discharges of biochemical oxygen demand (BOD) and ammonia to assure that the water quality standards for dissolved oxygen are achieved for a full range of hydrologic and climatic conditions. Simple and sophisticated models, ranging from desk-top calculations to computer programs that require considerable computational power have been used for these WLAs, with the selection of the required sophistication of the model typically being based upon the complexity of the surface water body being modeled and the potential impacts on the point sources that will be regulated as a result of the WLA. Data limitations for model calibration and verification also strongly influence the choice of a modeling approach. When large numbers of point sources are affected by a WLA and/or the resulting permit limits represent advanced treatment requirements, sophisticated and expensive modeling should usually be performed.

Waste load allocations are necessary when there is a documented or potential exceedance of a surface water quality standard. Waste load allocations result in water quality-based NPDES permit limits for point sources of the pollutant that is being allocated, in accordance with 40 CFR 122.44. By definition, these water quality-based permit limits are more restrictive than technology-based permit limits (because the WLA would be unnecessary if the technology-based limits achieved the water quality standard), and therefore point sources are required to reduce their loadings of the pollutant beyond that typically required of a similar point source that is not subject to water quality-based limits, e.g., categorical effluent limitations guidelines. Because the additional waste treatment or pollution prevention required by water quality-based permit limits can be very expensive, depending upon the extent of effluent loading reduction required to meet the limits, WLAs often are subject to intense public scrutiny. The more restrictive that the water quality-based limits resulting from the WLA are, the more closely the WLA assumptions and predictions will be studied by the regulated community and the interested public.

The principal steps in developing the WLA for any pollutant and any receiving water body are as follows:

1. Determine that a reasonable potential for exceedance of a water quality standard exists. This will usually be accomplished with historic water quality monitoring data; however, in some instances simple mass balances of point source waste loadings may be used to determine that the potential to exceed a standard exists.
2. Select the modeling approach that will be used to develop the WLA.
3. Collect the ambient surface water, nonpoint source, and point source data required by the model to simulate the receiving water concentrations of the target pollutant(s).
4. Calibrate and verify the model with ambient water quality data, sediment quality data, nonpoint source data, and point sourcc data for the pollutant(s) that will be allocated.
5. Use the model to calculate the allowable loadings of the pollutant that will assure that the water quality standard is achieved for the critical hydrologic condition (a low flow with specified return frequency for streams; selected tidal conditions for estuaries). The allowable point source loadings are usually constrained by a safety/growth factor (selected at the discretion of the regulatory agency), by the upstream (background) pollutant loading, and by the nonpoint source contribution.
6. Select a method to allocate the available loading capacity to each point source discharger.
7. Incorporate the allocations in each point source discharger's NPDES permit as numeric limits. Require best management practices (BMPs) for nonpoint sources as feasible.

The allocation of available loading capacity to point sources (after adjustment for upstream loadings and nonpoint source loadings) can be done by a number of methods. In its *Technical Support Document for Water Quality-based Toxics Control* (TSD) [1], the EPA lists 19 different waste load allocation methods. In practice, the most common approaches used are: (1) the equal concentration limit approach, and (2) the equal percent reduction in existing effluent loading for all sources of the pollutant. The equal concentration limit approach is generally considered to be more equitable to all point sources, because then only those sources that are above the WLA concentration limit, and are thus assumed to be contributing the most to the exceedance of the water quality standard, are required to reduce their effluent loadings.

It should be emphasized that this WLA approach is based on achieving a water quality standard during dry-weather conditions. Waste load allocations for the transient discharges associated with wet weather are not currently state-of-the-art. In fact, the appropriate water quality standards for toxic pollutants, including metals, which should apply for transient, wet weather flows have not yet been satisfactorily defined.

The conceptual approaches to waste load allocation modeling for metals are not unlike those for dissolved oxygen. However, the physical and chemical interactions between metals and the surface waters are very different than those considered in dissolved oxygen WLAs. They are not, however, significantly more complex than the biological, chemical, and physical relationships that control dissolved oxygen in surface waters.

Figure 1 is a simplified diagram of the physical and chemical phenomena that affect metals concentrations in surface waters. Metals can be present in the water column in either the dissolved or particulate form. The relative amounts of metal in each form are described by the partitioning coefficient, which is a function of the specific metal and its valence, the suspended solids concentration in the water column, the concentrations of other cations and anions, and pH. The EPA and most states now accept that the dissolved metals fraction in the water column is the form of metal that is most closely correlated to the toxicity of the metal to aquatic life. This is because it approximates the bioavailable form of the metal, although this is an oversimplification. Metals may be measured in the dissolved form (which is actually an approximation because it is based on the filtrate through a 0.4 micron filter) but actually be present as soluble complexes that are not bioavailable. Because surface water quality standards are either currently based on the dissolved fraction of metals (for example, those of Texas, Louisiana,

Arizona, and the National Toxics Rule), or probably soon will be, it is necessary to account for the partitioning of metals in WLA studies.

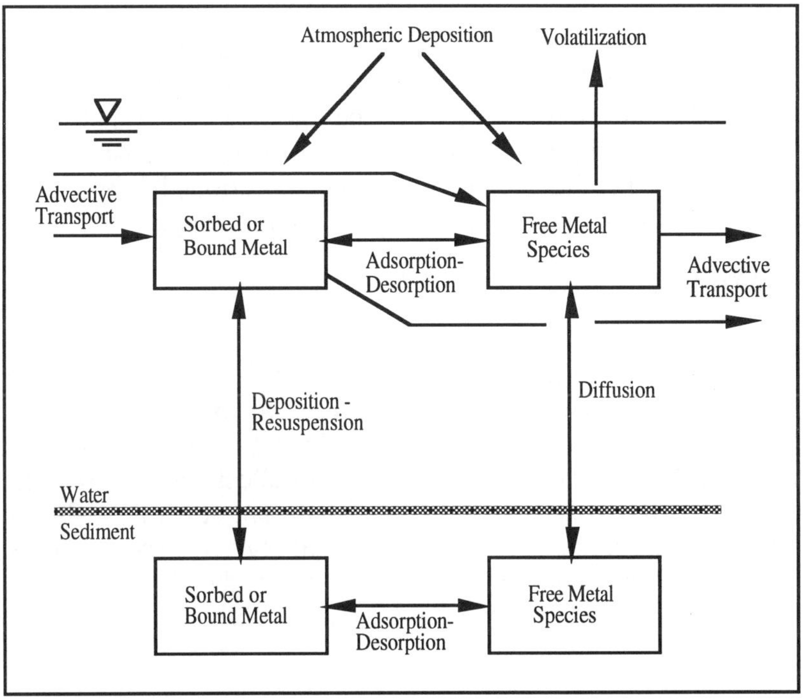

Figure 1. Transport and Fate Mechanisms.

Dissolved and particulate metals are transported by advective flow (stream flow, tidal motion) in surface water bodies. Therefore, the WLA model must provide an acceptable representation of the hydraulics of the surface water body being studied. Streams and rivers are the simplest system to model from a hydraulic standpoint, especially when a steady-state assumption is used for the critical hydrologic condition. The hydraulics of estuaries and lakes are more complex, but models are available that will provide satisfactory simulation of these water bodies if sufficient physical data are available to properly describe the system being modeled.

Particulate metals will settle to the bottom of a water body in locations where the velocity of the water is sufficiently low compared to the sedimentation velocity of the particles containing the metals. These particulate metals will then accumulate in the sediment. If the velocity in the water body increases sufficiently, for example during high streamflows, the particulate metals may be resuspended in the water column by a process that is called *scouring*. Particulate metals that are captured in the sediment can dissolve into the water that is present in the interstices between sediment particles, if the chemical conditions are right. These dissolved metals in the sediment interstitial water can then be bioavailable and potentially toxic to benthic organisms.

Sediment metals may also undergo biological transformation in the sediments. Mercury is the most widely-recognized metal in this regard. Inorganic mercury in sediments can be biologically transformed into organic mercury compounds, which are bioaccumulated by organisms higher on the food chain and can reach concentrations in commercial and game fishes and crustaceans that are toxic to humans.

2.1 Models

Although this brief description indicates the complexity of the behavior of metals in aquatic systems, most of these fate pathways for metals are reasonably well-understood and can be reliably simulated in a WLA model if sufficient effort is made to collect the site-specific data needed to describe these interactions. There are several public domain models that are available for modeling the fate of metals in surface waters. The EPA lists four such models in the *Technical Support Document for Water Quality-based Toxics Control* [1]. These models are:

- *DYNTOX* - a probabilistic dilution model for rivers;
- *WASP4* - a model that can be used in either steady-state or dynamic mode to model rivers, lakes, and estuaries;
- *HSPF* - a dynamic model for rivers; and
- *MINTEQA2* - a steady-state model for calculating equilibrium metals speciation in rivers, lakes, and estuaries.

In addition to these models, which have specific functionality for metals, metals can be simulated as conservative substances using any stream or estuarine water quality model that can simulate the behavior of conservative substances. For example, the widely-used QUAL-II model can be used to simulate metals concentrations in streams and rivers if the conservative assumption is used.

2.2 Modeling Approaches

Waste load allocation methods for metals, which use the above models or their equivalent, can be divided into the two generic approaches identified in the EPA's TSD:

- Steady-state modeling techniques; or
- Dynamic modeling techniques.

The simplest and most widely-used waste load allocation approaches are based on steady-state modeling. In the steady-state model a single set of hydrologic conditions (typically the critical low stream flow) is used for the simulation. Higher stream flows are used for model calibration and verification when ambient water quality data are available. The approach most often used for metals is to assume that they are conservative in the water column and that there is no change in the aqueous-solid phase partitioning relationship within a surface water segment. The next level of steady-state modeling complexity may include modeling of the aqueous-solid phase partitioning and/or sediment interactions with the water column. The level of complexity built into a model should be based on the degree of understanding of the physical and chemical relationships being simulated, for the particular body of water being studied. Thus, substantial amounts of site-specific physical, hydrologic, and chemical data are necessary to properly develop and apply a complex water quality model, even when it is based on steady-state hydraulics.

The steady-state model may be either overconservative or underconservative depending upon the assumptions made in its development and application. Generally, because the assumptions made for steady-state modeling are intentionally conservative, the resulting wastc load allocations are likewise conservative. Typically, steady-state WLAs are performed at the critical low flow condition in streams (the 7-day, once in 10 year recurrence interval low flow - 7Q10) and assuming that all point source dischargers are discharging at their maximum 30-day average flow rate and maximum allowable concentration of the pollutant to be regulated. Corresponding critical hydraulic conditions may be used for estuaries, which will be when there is minimum advective stream flow and minimum tidal exchange. Section 4.5.1 of the TSD describes steady-state modeling techniques that are applicable to the modeling of toxic pollutants, including metals.

Dynamic modeling allows the waste load allocation to be based on actual hydraulic records for the surface water body and expected variabilities in the source discharges of metals, based on actual source data. Variations of the dynamic modeling approach can include use of steady-state hydraulics but variable loadings from point and nonpoint sources based on the probabilities of occurrence of these loadings. In Section 4.5.1 of the TSD the EPA describes three approaches to dynamic modeling for WLAs:

- *Continuous simulation* - in this type of model, a continuous record of stream flow, point and nonpoint source effluent flows, and the corresponding pollutant concentrations including background concentrations are simulated on a day-by-day basis for the entire period of record available. The output of such a model is a continuous record of the concentrations of the pollutants of interest at all locations in the surface water body. These concentrations can be expressed as a probability distribution which can be used to determine the maximum allowable daily loading that will assure that the water quality criteria for the pollutant of concern can be met. The continuous simulation model obviously requires an extensive time series database for both the surface water and the effluent discharges. This requirement is usually the most significant limitation on its use in waste load allocations, because rarely are there extensive records with good data quality for metals in surface waters and effluents.

  A good example of continuous simulation modeling is a recently published paper on nutrient concentrations, algae growth, and dissolved oxygen in Chesapeake Bay. A calibrated hydrodynamic and water quality model of Chesapeake Bay is used to simulate algae growth and dissolved oxygen concentrations in response to multiple years of tributary inflows and point source discharge data [2].

- *Monte Carlo simulation models* - The Monte Carlo method of simulation combines a deterministic fate and transport model with probabilistic inputs for variables such as stream flow, tides, and point source loadings. Either a steady-state or dynamic model can be used. The model is typically run for a large number of randomly combined input conditions describing the full range of potential variation and expected frequency of occurrence for each variable. The input data are formatted to be consistent with the averaging period for the water quality standard that is to be achieved. For example, if the acute criterion is to be achieved at the 1-day average, 1 in 5 year low stream flow, then the frequency distributions of daily stream flow and daily effluent discharges from point sources are used as input to the deterministic model. If a 7-day average low flow is the target, then the distribution of 7-day average stream flows and effluent discharges are needed. Thus, some data manipulation is often needed to organize the data for the Monte Carlo simulation. The simulation model generates an acceptable waste loading that will achieve the target water quality for each set of input conditions simulated, and these waste loadings can then be ranked and the frequency of occurrence of each loading can be calculated (i.e., a frequency distribution can be obtained). The regulatory authority can then select the allowable waste loading with any specified probability of occurrence (e.g., the allowable total waste loading that assures that the water quality standard will be met 95% of the time).

The advantages of the Monte Carlo analysis are that it can take into account fate and transport processes in surface waters, it does not require time series data although it does require sufficient data for each variable to establish reliable frequency distributions, and it predicts a frequency distribution of surface water quality which allows a waste load allocation to be based on a predicted frequency of occurrence. Its biggest disadvantage is the requirement for reliable distributions of metals loadings in point source and nonpoint source discharges and the surface water.

- *Lognormal probabilistic dilution model* - With this approach, the frequency distributions of stream flow and metals concentration are mathematically combined with the corresponding flows and concentrations for all point and nonpoint sources. This approach assumes complete mixing of the stream and all effluents and does not account for fate and transport of the metal - it is strictly a mass balance. The output is a frequency distribution of surface water concentrations of the target metal, which can be used to establish the waste load allocation that will assure that the water quality standard for the metal is achieved at a specified frequency of occurrence. The biggest advantage of this approach is that it is relatively simple if only one or a few point sources of the metal must be analyzed. Its disadvantages are that it does not account for transport and fate of the metal, all input distributions must be assumed to be lognormal, and it is difficult to apply and is inaccurate when there are a significant number of sources of metals for a specific surface water body.

A hybrid form of dynamic:steady-state WLA modeling uses only the expected variations in the point source discharges, and the critical low-flow or tidal dispersion hydrodynamics, for the TMDL calculation. This approach takes into account the fact that the probability that all point sources will be discharging at their maximum monthly average loadings at the same time that the surface water body is at the critical hydraulic conditions for dilution and transport is probablistically remote, to the extent that it is practically impossible when there are a significant number (>5) of point sources. In this approach, the WLA is performed in the conventional way (assuming that all point sources discharge at their maximum monthly average loading), and then the resulting TMDL limits are then adjusted upward a uniform amount to reflect the unlikely probability that all such maximum loadings will occur at the same time and be coincident with the critical hydrologic conditions. Because monthly average effluent loadings are typically assumed to follow a normal statistical distribution, and such distributions are additive, this calculation is relatively straightforward.

Dynamic models have substantially greater data requirements for waste load allocations than steady-state models. Reliable long-term historic records of point source discharges, stream flows, and background metals concentrations are required for dynamic models. An example of the application of dynamic modeling to waste load allocations for metals was recently published [3]. Continuous simulation techniques were used to model three rivers in Texas. Stream flow, hardness, and suspended solids were used to calculate permissible zinc loadings to the three rivers. The model included partitioning between the dissolved and particulate phases, but did not simulate zinc speciation. The study concluded that dynamic modeling demonstrated that allowable waste load allocations were two or more times greater than those predicted by steady-state, critical condition, models. This difference is attributable to the fact that critical stream flows, minimum suspended solids concentrations, and minimum hardness concentrations are not coincidental, as is assumed by the steady-state, critical condition approach.

A model that accounts for the dynamics of surface water and effluent quality to the extent feasible, given available time and resources, should be the preferred approach for waste load allocations involving large numbers of sources. Because the potential costs of achieving the TMDLs in surface waters that have many point and nonpoint sources of metals may be very high, every effort should be made to assure that the waste load allocation is as accurate as practical and not overly conservative.

2.3 Calibration and Verification

Regardless of the modeling approach or type of model used for a WLA, a critical element of the analysis is the calibration and verification of the model. Calibration and verification refer to using the model to simulate historic water quality conditions, and adjusting the physical and chemical variables in the model until acceptable accuracy is achieved for a range of hydrologic conditions. The term calibration is typically used to describe the adjustment of the various coefficients in the model (sedimentation rates, partitioning coefficients, resuspension rates, etc.) using one or more historic hydrologic and water quality data sets. Verification involves using the calibrated model to simulate one or more additional data sets that were not used for calibration, to determine how well the calibrated model can predict the measured water quality.

Calibration and verification is an iterative process, and some modelers do not make a distinction between these two steps. At any rate, only through the calibration and verification process can the modeler be assured that predicted water quality and the resulting WLA are sufficiently accurate to allow protective and equitable water quality-based permit limits. Obtaining the necessary hydrologic and water quality data for proper calibration and verification of a water quality model is typically the most expensive portion of a WLA.

Determining the goodness-of-fit of a water quality model to ambient data can be done graphically, statistically, or using a combination of the two methods. One very simple approach that combines graphics and statistics is to use a linear correlation between measured ambient and model-predicted pollutant concentrations. The goodness-of-fit can be determined by the slope of the regression line with the ambient concentrations being used as the independent variable and the predicted concentrations being used as the dependent variable. If the slope of this line is 1.0 and the intercept is at zero, then there is perfect correlation and the model predictions are exactly the same as the actual measurements. A slope greater than 1.0 means that the model overpredicts the actual concentrations; a slope less than one means that the model underpredicts the ambient concentrations. The regression coefficient (R^2) indicates the precision of the model, i.e., how much of the variance in the measured concentrations are explained by the model predictions.

A commonly-used graphical approach is to plot the predicted concentrations and measured concentrations for a specific set of hydrologic conditions as a function of location in the water body being simulated. If multiple measurements of concentration at a single site are available for the hydrologic period, then the range of measured concentrations are shown on the plot. This approach provides a visual display of the model's predictive capacity at each point within the water body, and allows identification of specific locations where the model may be inaccurate.

Other statistical approaches are also available, including, for example, the calculation of confidence limits on the simulation results using measured data to establish the confidence levels. This approach is an excellent way to provide the public and regulated community with an understanding of how much confidence can be placed in the WLA results.

The need for calibration and verification of a WLA model cannot be overemphasized when the results of the WLA will be used to establish TMDLs that will be much more restrictive than technology-based limits. For metals, this will be the case when the TMDLs result in water quality-based permit limits below concentrations of 20-50 μg/L, depending on the metal being regulated. At these low concentrations, conventional metals removal technologies (chemical precipitation) will often not provide sufficient treatment, and more exotic and expensive processes (reverse osmosis, ion exchange, multi-stage chemical precipitation followed by filtration) must be applied to achieve the TMDL target. Because the compliance costs will be high in such cases, it is incumbent on the agency conducting the WLA study to maximize the accuracy and precision of the simulations. This will require sufficient ambient water quality and hydrologic data to properly calibrate and verify the simulation model that is used for the WLA, and presentation of the precision and accuracy of the model predictions.

3. CASE HISTORIES

The best way to describe WLAs for metals is by use of case history data for completed studies. Examples of waste load allocations for metals include: Pennsylvania's simplified mass balance modeling, the Texas modeling of nickel for the Houston Ship Channel, EPA's Flint River Study, and the recently-completed New York Harbor Study. These case histories represent a range of modeling approaches, from very simple mass balances to highly complex simulations. All, however, are based on steady-state hydrology (i.e., critical low flow).

3.1 Pennsylvania Watershed Analyses

The Pennsylvania Department of Environmental Resources (DER) performs a stream segment WLA for toxic metals for every NPDES permit issuance. This analysis is not based on a prior determination that the water quality criterion for a specific metal is potentially exceeded; it is performed in every basin to provide a check on the available dilution for each point source and each metal with a standard. The DER uses a simple mass-balance-dilution model, assuming a 7Q10 critical dilution flow, for this WLA. No fate mechanisms, such as sedimentation, are included in this modeling. The metals are modeled as total recoverable metals concentrations and therefore bioavailability of the metals are not considered. All point sources of a metal in the stream segment are assumed to be discharging at their maximum monthly average flow rate and a maximum monthly average effluent concentration of the metal. Background (natural and nonpoint source) concentrations of each metal simulated are considered when data are available. The models for each basin rely on site-specific hydrology and stream physical conditions, but no calibration and verification of the models with ambient water quality data are performed.

The model is used to demonstrate that the total mass of a metal discharged to a specific stream segment does not exceed the applicable chronic water quality criterion at any location within the stream segment. If the water quality criterion for a metal is determined to be exceeded due to the discharge of any point source, the allowable discharge of the metal is decreased until the standard is achieved. This is done by requiring an equal percent reduction of the metal discharged by all point sources. The resulting discharge allowance then becomes the NPDES permit limit for that discharger.

This approach to WLA is very conservative and is useful to assure that a standard dilution analysis does not result in an exceedance of a water quality criterion because of the cumulative effect of multiple point source discharges. Because of the method's simplicity, it can be used on every stream in the state without requiring extensive and expensive water quality studies. However, the conservative nature of this approach assures that it will be overprotective and will inevitably result in water quality-based limits for metals that are not as cost-effective as limits based on a more comprehensive analysis. For some dischargers, this approach could result in unachievable water quality-based permit limits. Pennsylvania does, however, have provisions in its water quality standards and permits that allow dischargers to develop site-specific water quality criteria that take into account aqueous-solid phase partitioning of metals, and bioavailability as determined by the water effects ratio (WER). These procedures offer a safety valve for any dischargers or group of dischargers that believe that the water quality-based limits which result from the WLA procedure are unnecessarily restrictive.

3.2 Houston Ship Channel/San Jacinto River

The Houston Ship Channel and San Jacinto River (tidal), in Harris County, Texas, are tidally-affected tributaries of Galveston Bay. The Houston Ship Channel area is one of the most industrialized surface water bodies in the U.S. This 40-kilometer (km) body of water receives over 300 million gallons per day (MGD) of treated municipal sewage and industrial wastewater from over 450 permitted point sources. The San Jacinto River, which joins with the Houston Ship Channel before they reach Galveston Bay, has only two major industrial dischargers and a number of small municipal treatment plants. There are four separate classified stream segments that make up the Houston Ship Channel-San Jacinto River system.

As a result of a joint water quality study of these waters by EPA Region 6 and the Texas Natural Resource Conservation Commission (TNRCC), formerly the Texas Water Commission), the EPA requested that a TMDL analysis for nickel be conducted by the TNRCC because of a potential exceedance of the chronic aquatic life-based water quality criterion in the Houston Ship Channel and San Jacinto River (tidal). (It should be noted that this EPA determination was based on one set of ambient surface water data that was not collected using clean techniques and which clearly reflected sample contamination during either collection or analysis or both.) The Texas chronic aquatic life criterion for nickel is 13.2 μg/L as dissolved nickel and the acute criterion is 119 μg/L, also as dissolved nickel. The TNRCC's default mixing zone analyses are such that the potential to exceed the chronic criterion drives the water quality-based permit limits and the acute criterion will never be the controlling factor. The TNRCC conducted a TMDL study to determine if the Agency's standard waste load allocation procedures that are used for determining water quality-based limits needed to be replaced by more restrictive segment-wide waste load allocations [4].

Effluent data for dissolved and total nickel were collected from all significant municipal and industrial dischargers under the authority of a Clean Water Act (CWA) Section 308 Order for Information. Each discharger was required to collect and analyze one sample of effluent per week for 4 weeks. The analyses were required to be performed with the most sensitive approved analytical method for nickel - graphite furnace atomic absorption. The dischargers also were required to report the effluent flows when each sample was collected.

The TNRCC used the QUAL-TX water quality model to simulate dissolved nickel as a conservative substance. The QUAL-TX model is a version of the QUAL-II model developed by the TNRCC and includes tidal dispersion effects as well as advective flows so that it can be used to simulate water quality in estuarine waters. QUAL-TX is a one-dimensional, steady-state hydraulics model. The critical low flow hydraulics for the Houston Ship Channel-San Jacinto River version of QUAL-TX have been calibrated and verified using salinity, dissolved oxygen, biochemical oxygen demand, and ammonia data. Because the model is calibrated for salinity, a conservative substance, it was also considered to be calibrated for dissolved nickel when this metal was simulated as a conservative substance.

The reason for assuming that dissolved nickel behaved as conservative substance was that this approach would result in the most restrictive TMDL under critical hydrologic conditions, which would then be compared to the water quality-based permit limits that dischargers would receive using the default TNRCC mixing zone and dilution factor. The TNRCC considered this approach as a screening WLA, which would be refined if the TMDL-based permit limits were more restrictive than the water quality-based permit limits that are based on the default dilution factor.

The QUAL-TX model was used to simulate the dissolved nickel concentrations under critical low freshwater flow conditions and critical (minimal) tidal dispersion. Background nickel concentrations were assumed to be 5 µg/L for all freshwater inflows and nonpoint source flows (load allocation). These background concentrations were deliberately assumed to be higher than was actually expected (there were no reliable background concentration data available at the time; a subsequent study has shown the true background nickel concentrations to be < 1 µg/L). Based on the effluent sampling data and the TNRCC default dilution factors, all municipal effluents were assigned an initial dissolved nickel concentration of 5 µg/L and industrial dischargers were assigned a discharge concentration of 30 µg/L. The model was then operated iteratively to reduce the allowable effluent concentrations in a segment if simulation predicted that the water quality criterion was exceeded.

The resulting TMDL analysis predicted an allowable dissolved nickel discharge criterion for each major municipal and industrial discharge. The draft TMDL has never been finalized, however, because the TNRCC determined that the default mixing zone and dilution factor, combined with the overall low amounts of dissolved nickel that were discharged from all point sources as demonstrated by the point source characterization, provided sufficient assurance that there was no reasonable potential for the nickel criterion to be exceeded. Three surveys of ambient water quality under low flow conditions that have been conducted since the draft TMDL was prepared, in which clean sampling and analytical methods were used for all metals, have demonstrated that the dissolved nickel concentrations in the Houston Ship Channel-San Jacinto River are actually substantially below the chronic aquatic life criterion. These studies have, in effect, verified the results of the WLA study which predicted that the standard was not jeopardized.

3.3 Flint River, Michigan

When the EPA developed its technical guidance manual for waste load allocations of toxic substances in streams and rivers [5], it performed a demonstration project for the development of procedures for regulating multiple point source discharges of toxic pollutants. This case study was for cadmium, copper, and zinc in a specific segment of the Flint River of Michigan, a major tributary to Saginaw Bay. The 60-km segment of the river modeled included the City of Flint and several major wastewater discharges.

The Flint River Study is a good case history of the data needed to develop a complex steady-state simulation model for metals. The model simulates suspended solids settling and resuspension, and dissolved and particulate concentrations of the three metals. Three comprehensive stream surveys, at low flow (August), a mid-range flow (December), and a high flow following a snow-melt event (March) were used to calibrate the MICHRIV water quality model for the three metals. Several different approaches to stream data collection were used, including: (1) daily grab samples at 13 river stations and 4 tributary stations, 24-hour composites at three point sources, and diurnal grab samples at 4 stream stations for a period of two weeks; (2) slug sampling (i.e., following a single parcel or *slug* of water downstream) using dye to follow the slug downstream and collecting samples from each tributary and point source as the slug passed it; and (3) sampling fewer river stations more frequently (every 4 hours) along with increased frequency of point source sampling (every 4 hours) and tributaries (every 12 hours) for a total survey period of 72 hours. Data collected included stream flows, dissolved and total metals, suspended solids, hardness, conductivity, pH, temperature, and dissolved oxygen.

The model was calibrated for each of the 3 survey periods. Suspended solids concentrations were calibrated first, using sedimentation rates and resuspension rates as the independent variables. After suspended solids concentrations were satisfactorily predicted for a survey period, calibration for each of the three metals was performed by adjusting the aqueous:particulate partitioning coefficients for each metal. The suspended solids calibration was not tampered with during the calibration for each metal.

The study achieved satisfactory calibration of dissolved and particulate cadmium, copper, and zinc concentrations for each survey period. The suspended solids settling rate and resuspension rates were significantly affected by stream flow because of the higher velocities that occurred as the stream flow increased. The partition coefficients were also affected by stream flow, because the larger particles that were maintained in suspension at the higher stream velocities contained proportionately lower concentrations of metals, and thus had lower partitioning coefficients.

This case study did not actually calculate TMDLs for the three metals. Instead, it was intended as an example on how to formulate and calibrate a steady-state waste load allocation model for metals. It is an excellent example of the effort required to develop a reliable model, and should be required reading for anyone who must develop such a model for a comprehensive TMDL analysis. The case study also shows how much field data is needed for satisfactory calibration of a steady-state model that reliably simulates the fate of metals in a stream or river. The costs of the necessary data collection, in addition to the efforts required to calibrate and validate the simulation model, are considerable.

3.4 New York/New Jersey Harbor

The New York/New Jersey (NY/NJ) Harbor is listed under Section 304(l) of the Clean Water Act as not attaining the water quality criteria for copper and mercury. In addition to these two metals, six other metals were identified as potentially exceeding their respective criteria. As required by the CWA, TMDLs had to be developed to allow allocating the acceptable loading capacity for each of these metals to point source dischargers that would assure attainment of the criteria. A joint WLA effort by the states of New Jersey and New York and Region II of the EPA was implemented to develop these TMDLs. The resulting TMDL methodology is one of the most complex and comprehensive efforts conducted to date to develop WLAs for metals (see Chapter 5, this volume). The report on the modeling of the harbor complex is an excellent source of information on how such a study should be conducted [6].

The water quality model used for this study was the Chemical Transport and Analysis Program (CTAP). This is a steady-state model that simulates the following fate processes:

- advective/dispersive transport;
- solid phase vertical transport;
- phase partitioning and kinetic reactions;
- transport across the air/water interface; and
- point and nonpoint source loadings.

Because of the complexity of the hydraulics of the NY/NJ harbor system, the model consisted of 490 water column segments and 396 sediment segments. The water column segments consist of 302 one-layer segments and 188 two-layer segments to simulate stratification. The physical transport used in the CTAP is simulated with a three-dimensional hydrodynamic model. The hydrodynamic model was calibrated for a low-flow period (October 1991) and a high flow period (January 1991) using USGS stream flow records for freshwater inflows from the tributaries and discharge monitoring report records for point source effluent discharges. Data were also collected on combined sewer overflows (CSO) for input into the model during wet weather simulations. Salinity was used to calibrate the advective and dispersive transport in the model.

Ambient data were collected for both the high flow and low flow calibration periods. Suspended solids data from these surveys were used to calibrate the model for solid phase vertical transport. The ambient metals data were used to calculate the site-specific partitioning coefficients for the model. Sediment metals partitioning data for the model were obtained from a 1992 sediment survey.

The model was calibrated for each of the eight metals that were simulated (copper, mercury, lead, cadmium, nickel, silver, arsenic, and zinc) using the January 1991 and October 1991 ambient data sets and the corresponding point source discharges, nonpoint source discharges, and tributary inflows. Calibration was successful using the tributary loading, nonpoint source loading, and point source data for 5 of the 8 metals.

The mercury loadings from the point sources and tributaries did not allow calibration of the model; ambient concentrations were consistently greater than predicted. A mercury loading approximately 9 times greater than the point source loadings had to be added to the system to achieve calibration. It was hypothesized that atmospheric deposition and/or methylation of mercury in the sediments were responsible for the added loading to the harbor waters.

Insufficient ambient data were available for arsenic to allow acceptable calibration of the model. Also, no CSO or wet weather influent data were available for arsenic.

Predicted concentrations of dissolved and particulate silver, and silver in the sediments, were all greater than the ambient data. A lack of data for total recoverable silver at key locations in the harbor and poor characterization of silver in CSOs and other waste discharges were believed to be the principal problems that prevented acceptable calibration. The difficulty of measuring accurate silver concentrations at trace levels in both the ambient water and effluents is suspected as being the major cause of the poor calibration for this metal.

The TMDL study included an analysis of the accuracy of the model predictions. The NY/NJ Harbor was divided up into regions and the historic ambient metals data were averaged in each region to compare to the model predictions for the same time periods. The 95% confidence limits for the prediction of regional data averages ranged from ± 10% to + 86%/-46%. The poor accuracy was for harbor areas for which there was a paucity of ambient and/or loading data.

The calibrated model was used to develop TMDLs for the NY/NJ Harbor under critical low flow conditions. TMDLs for the harbor can be expressed on a system-wide, regional, or segment-specific basis. Because of the complexity of the transport patterns in the harbor, a number of different WLAs are possible for a given TMDL. To make the development of the WLAs more efficient, the model results were provided in spreadsheet format. The regulatory agencies can use this spreadsheet to allocate required reductions to all point/nonpoint sources within a given segment of the harbor.

The NY/NJ Harbor TMDL study is a good example of the massive amount of data collection that is required to achieve a calibrated model of a complex surface water system. This TMDL effort should be studied by anyone who is planning a TMDL study of tidally-influenced system which receives a large volume and number of wastewater discharges that will be affected by the modeling results. It represents the current state-of-the-art for metals WLA studies.

4. CONCLUSIONS

As discussed in this paper and demonstrated by the examples presented, WLAs for metals are not substantially different in either concept or complexity than WLAs for dissolved oxygen. The modeling of surface water system hydrodynamics and advective and dispersive transport phenomena are no different for metals than they are for any other chemical. The major fate pathways for metals - sedimentation, solids resuspension, equilibrium partitioning, advective transport - are reasonably well-understood and theoretical and empirical relationships that can be used in simulation models are available. The principal differences between modeling dissolved oxygen and the fate of metals in surface water systems are that the ambient surface water, sediment, and related data that are required to simulate the biological, physical, and chemical processes that influence metals are not generally available for most surface waters. One major problem is that most historic ambient metals databases are of suspect validity because of sample contamination during collection and analysis and the resulting concentration data have a high bias compared to true ambient concentrations (see Chapter 6, this volume). Even if sample contamination were not a problem, a great deal of the historic ambient metals data were measured using analytical methods which are insufficiently sensitive to detect the trace concentration levels represented by actual ambient concentrations and the surface water quality standards (e.g., copper, lead, cadmium, silver). Historic metals data from marine systems were not measured with analytical methods that remove the interference from high concentrations of the common sea salts, which also interfere with trace metals analysis for many metals.

Thus, a considerable amount of new ambient metals data is required for many surface waters before a determination can even be made as to whether or not there is a reasonable potential that a surface water quality criterion is exceeded or could be exceeded. The data required to calibrate a metals WLA model for a specific surface water are even greater than that needed to make a criteria attainment determination, as demonstrated in the Flint River and NY/NJ Harbor studies.

Unfortunately, the cost of obtaining reliable data for trace metals in ambient waters is considerably greater than that for obtaining dissolved oxygen data and the related parameters. If clean sampling and analytical techniques and analytical methods with maximum sensitivity are not used, the ambient metals data will be suspect and thus worthless. This means that calibrated WLA models for metals will probably be the exception rather than the rule.

Because of the difficulty and expense of collecting the data and developing calibrated WLA models for metals, the need for calibrated WLA models should be determined based on the expected economic impacts of the TMDLs. For bodies of water like the NY/NJ Harbor, substantial decreases in the allowable permit limits for some metals (copper, mercury) would potentially cost municipal and industrial point source dischargers hundreds of millions, and perhaps billions, of dollars. If a properly-conducted WLA determines that point sources contribute negligibly to a criteria exceedance (e.g., mercury in NY/NJ Harbor), then the money that would be spent to reduce point source discharges can more effectively be spent to reduce the principal source contributions or otherwise mitigate its effects. In such a context, the time and cost involved in developing an adequately calibrated model of the surface water system is an intelligent use of resources.

The WLA modeling approach that is selected for a particular study should be based on the objective of the study, the complexity of the surface water system (including the number and type of sources), and the potential impacts of the TMDL results. As shown by the case histories presented here, if simple screening of effluents to assure that water quality criteria are achieved when default mixing zone procedures are used to set permit limits is the objective, simple mass-balance and steady-state modeling approaches are cost-effective and adequate. However, once it is apparent that TMDLs will require substantial reductions in point source and nonpoint source loadings beyond those required by default mixing zones, then more sophisticated modeling is needed.

The author believes that for complex surface water systems, some form of dynamic modeling of metals concentrations in water and sediment should be the ultimate objective for TMDL development. Because the data requirements for dynamic models are extremely large, and multiple years of reliable records for natural and man-made sources of metals to the surface water are required, it will be a considerable time in the future until such models are practical because of the reliability problems with historic metals data. However, regulatory agencies should start now to develop the necessary ambient surface water data as part of their ongoing water quality monitoring programs. This means the collection of reliable metals data using clean sampling and analytical methods, use of the most sensitive analytical methods available for all metals analyses, collecting dissolved and particulate metals data for every ambient surface water sample, measuring suspended solids in every metals sample, and characterizing the metals concentrations, particle size, and acid volatile sulfide concentrations in sediments.

REFERENCES

1. U.S. EPA, "Technical Support Document for Water Quality-based Toxics Control," EPA/502/2-90-001, Office of Water, March (Washington, DC 1991).

2. Cerco, C.F., "Simulation of Long-Term Trends in Chesapeake Bay Eutrophication," *J. Env. Eng.* 21:298-310 (1995).

3. Parkerton, T.F., S.M. Stewart, K.L. Dickson, J.H. Rodgers and F.Y. Saleh, "Derivation of Site-specific Water Quality Criteria for Zinc: Implications for Wasteload Allocation," *J. Water Pollut. Control Fed.* 61:1636-1644 (1989).

4. Texas Water Commission, "Total Maximum Daily Load for Dissolved Nickel in the San Jacinto River Tidal, Houston Ship Channel, and Buffalo Bayou," Draft, November (1991).

5. Delos, C.G., W.L. Richardson, J.V. DePinto, R.B. Ambrose, P.W. Rogers, K. Rygwelski, J.P. St. John, W.J. Shaughnessy, T.A. Faha and W.N. Christie, "Technical Guidance Manual for Performing Waste Load Allocations," in *Book II Streams and Rivers, Chapter 3 Toxic Substances*, EPA-440/4-84-022, Office of Water Regulations and Standards, June (Washington, DC 1984).

6. HydroQual, Inc., "Development of Total Maximum Daily Loads and Wasteload Allocations for Toxic Metals in NY/NJ Harbor, Modeling Report," Prepared for U.S. EPA Region II, New York/New Jersey Harbor Estuary Program, Water Management Division, Surface Water Branch, September (New York, NY 1995).

5

DEVELOPMENT OF SITE-SPECIFIC MARINE COPPER CRITERIA FOR NEW YORK/NEW JERSEY HARBOR

Felix Locicero,[1] Wayne F. Jackson,[1] Patrick M. Pergola,[1] Thomas M. Brosnan,[2] David J. Hansen,[3] Glen B. Thursby,[3] Herbert E. Allen[4] and Dion A. Lewis[5]

[1]U.S. Environmental Protection Agency
Region II, Water Management Division
New York, NY 10007

[2]New York City Department of Environmental Protection
Marine Science Section
Wards Island, NY 10035

[3]U.S. Environmental Protection Agency
Environmental Research Laboratory
Narragansett, RI 02882

[4]Department of Civil and Environmental Engineering
University of Delaware
Newark, DE 19716

[5]Battelle Ocean Sciences
Duxbury, MA 02332

1. INTRODUCTION

Section 304(l) of the Clean Water Act required, in part, that states develop and submit to the EPA a list of waters impaired entirely or substantially due to point source dischargers. A review of the historic ambient water quality and discharger loading data indicated that the waters of the New York/New Jersey Harbor were water quality-limited for copper and mercury. As a result, the states of New York and New Jersey listed the waters of the Harbor (Figure 1) as water quality-limited for copper and mercury, due entirely or substantially to point source discharges.

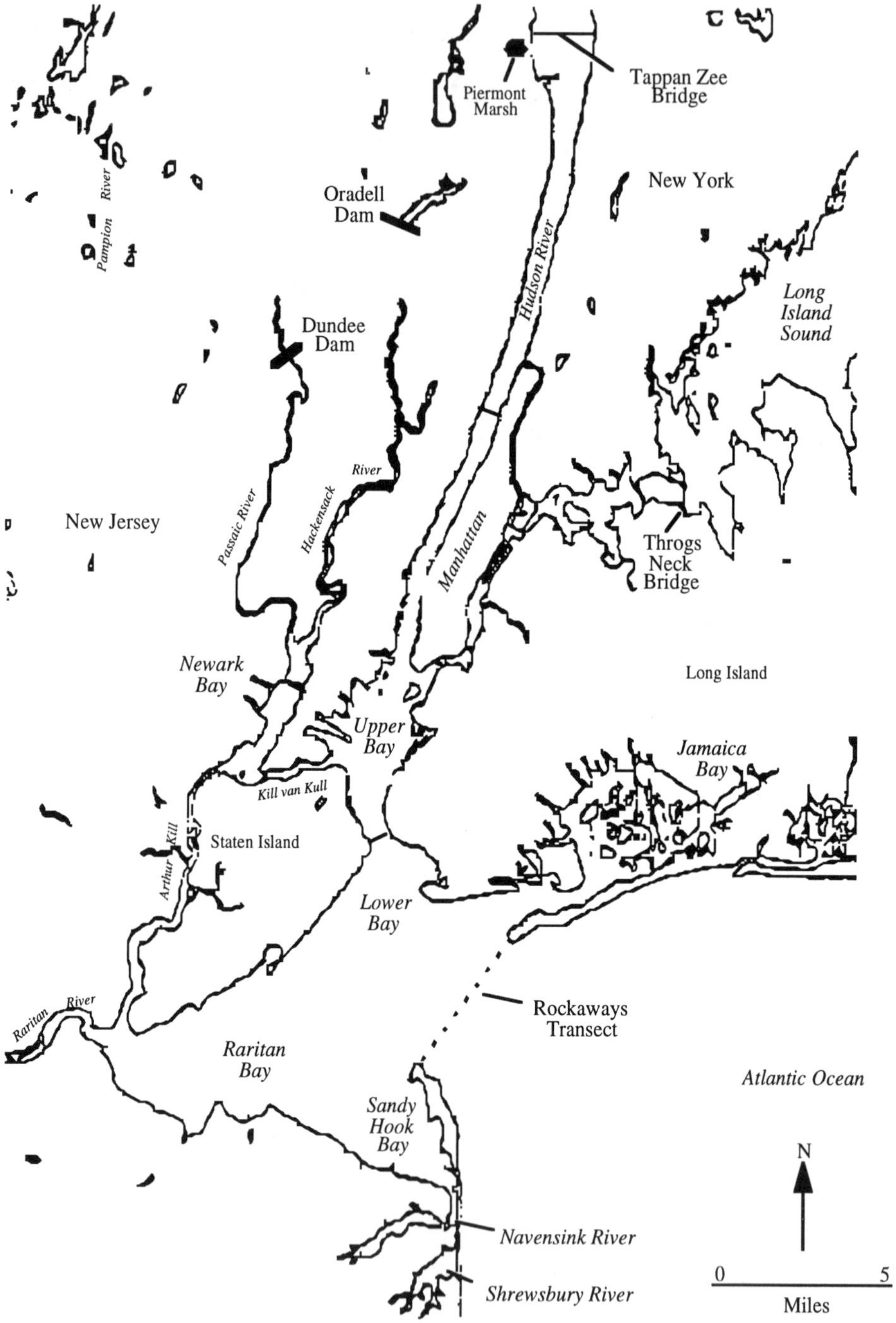

Figure 1. ***New York/New Jersey Harbor Estuary Project Area.***

The waters of the Harbor include: the Hudson River (to the Tappan Zee Bridge), the Harlem River, the East River, the Upper Bay, Newark Bay, the Passaic River (to the Dundee Dam), the Hackensack River (to the Ordell Dam), the Kill van Kull, the Arthur Kill, Raritan Bay, the Raritan River (to the Fieldsvill Dam), the Lower Bay to the Rockaway-Sandy Hook Transect) and Jamacia Bay. In order to develop total maximum daily loads/waste load allocations/load allocations (TMDLs/WLAsLAs), a Water Quality Standards/Total Maximum Daily Load Workgroup (the Workgroup) was formed under the auspices of the New York/New Jersey Harbor Estuary Program (HEP). The Workgroup was charged with the identification of metals of concern, review and selection of a set of uniform metal criteria to be applied by the states of New York and New Jersey, and to develop TMDLs/WLAs/LAs, as necessary.

The Workgroup included representatives of the EPA, the NYSDEC, the NJDEP, the NYCDEP, the ISC, the CAC, the STAC and local municipal and industrial dischargers.

2. AMBIENT DATA

A review of historic ambient metals data from several agencies indicated that the waters of the Harbor were exceeding the applicable total recoverable based criteria for copper, mercury, nickel, lead, arsenic, cadmium, silver and zinc [2]. However, ambient water quality analyses conducted in the Harbor during 1988 by Battelle Ocean Sciences using "ultra clean techniques" and the acid soluble approach indicated that the ambient concentrations of metals in the Harbor were much lower than historical data indicated. This raised two issues: the validity of the historic metals data and the appropriate species of metal to control. These two issues, we discovered, were of national concern.

In order to address these issues, extensive Harbor-wide ambient and effluent metals monitoring was conducted during the 1990-1992 period [3-5]. The Battelle data collection demonstrated that only copper, mercury, nickel and lead exceeded or potentially exceeded applicable total recoverable based water quality standards in all or part of the Harbor. The Workgroup chose to use the most stringent state water quality standard for mercury, nickel and lead, expressed as the total recoverable form of the metals, to implement TMDLs. However, for copper, the Workgroup proposed development of the Harbor Site-Specific Marine Copper Criteria. The development of the Site-Specific Marine Copper Criteria is discussed in detail in "Development of Site-

Specific Copper Criteria for the NY/NJ Harbor Complex Using the Indicator Species Procedure" [6].

TMDLs/WLAs/LAs were developed for copper, mercury, nickel and lead. The development of the TMDLs/WLAs/LAs is discussed in "Total Maximum Daily Loads (TMDLs) for Copper, Mercury, Nickel and Lead in NY-NJ Harbor" [7]. The remainder of this paper will focus on the development of the Harbor Site-Specific Marine Copper Criteria.

3. HARBOR SITE-SPECIFIC MARINE COPPER CRITERIA STUDY

The Workgroup undertook the development of the Harbor Site-specific Marine Copper Criteria for the following reasons:

1) NYSDEC was in the process of adopting a dissolved copper standard and NJDEP's copper standard was total recoverable based;
2) The ambient copper data collected by Battelle [3-5] indicated that the total recoverable form of copper ranged between approximately 3.0 to 50.0 μg/L. It was demonstrated that as total suspended solids (TSS) increased, the concentration of total recoverable copper also increased. The dissolved form of copper ranged from about 0.33 to 3.05 μg/L and did not seem to increase with increasing TSS concentrations;
3) Available ambient toxicity tests indicated that metals based toxicity was not occurring in the Harbor;
4) The implementation of the most stringent standard for copper would have a significant compliance cost, if the WLAs were even technically achievable; and
5) EPA headquarters was beginning to suggest that criteria for metals be expressed as the dissolved form.

4. DESIGN

The study design was based on the use of Indicator Species [8] and the Recalculation Procedure [1] approach. EPA guidance on the Indicator Species Procedure was limited in terms of application of this procedure in an estuary, especially at a site as complex as the New York/New Jersey Harbor. The study was designed in cooperation with the EPA's Environmental Research Laboratory - Narragansett (ERL-N) and the Workgroup.

The Indicator Species approach is used where it is believed that the physical/chemical characteristics of the ambient site water may influence the biological availability and/or toxicity of a specific pollutant. The Indicator Species approach uses toxicity tests conducted in both site and control water to develop a biologically-based adjustment to a numeric criteria. This biologically-based adjustment is known as a "Water Effect Ratio" (WER). A WER is the product of the toxicity endpoints (geometric mean of LC/EC50s) of site water divided by the toxicity endpoints (geometric mean of LC/EC50s) of control water. The final adjusted criteria is the product of the WER multiplied by the appropriate numeric criteria.

Seven sites (Figure 2) were chosen to represent the major hydrographic areas of the Harbor and to maximize water collection over the entire range of observed TSS levels (1.28 to 409 µg/L). Two sampling events were scheduled, one at high and one at low flow conditions. The low flow sampling event was conducted in October 1992, and the high flow event was conducted in April/May 1993.

Water from Narragansett Bay, Rhode Island was used as the "laboratory" control water for this study. Narragansett Bay was chosen because it is dilution water typically used to develop national marine water quality criteria. The Narragansett Bay water was sand-filtered.

A single water sample was collected from six of the seven Harbor sites. Three water samples were collected over the tidal cycle at the Hudson River (H3) site in order to maximize the chance of recovering high TSS levels.

The following toxicity testing species were chosen for: their sensitivity to copper; the availability of standard test methods (either ASTM or EPA); and the contract laboratory's ability to conduct the tests.

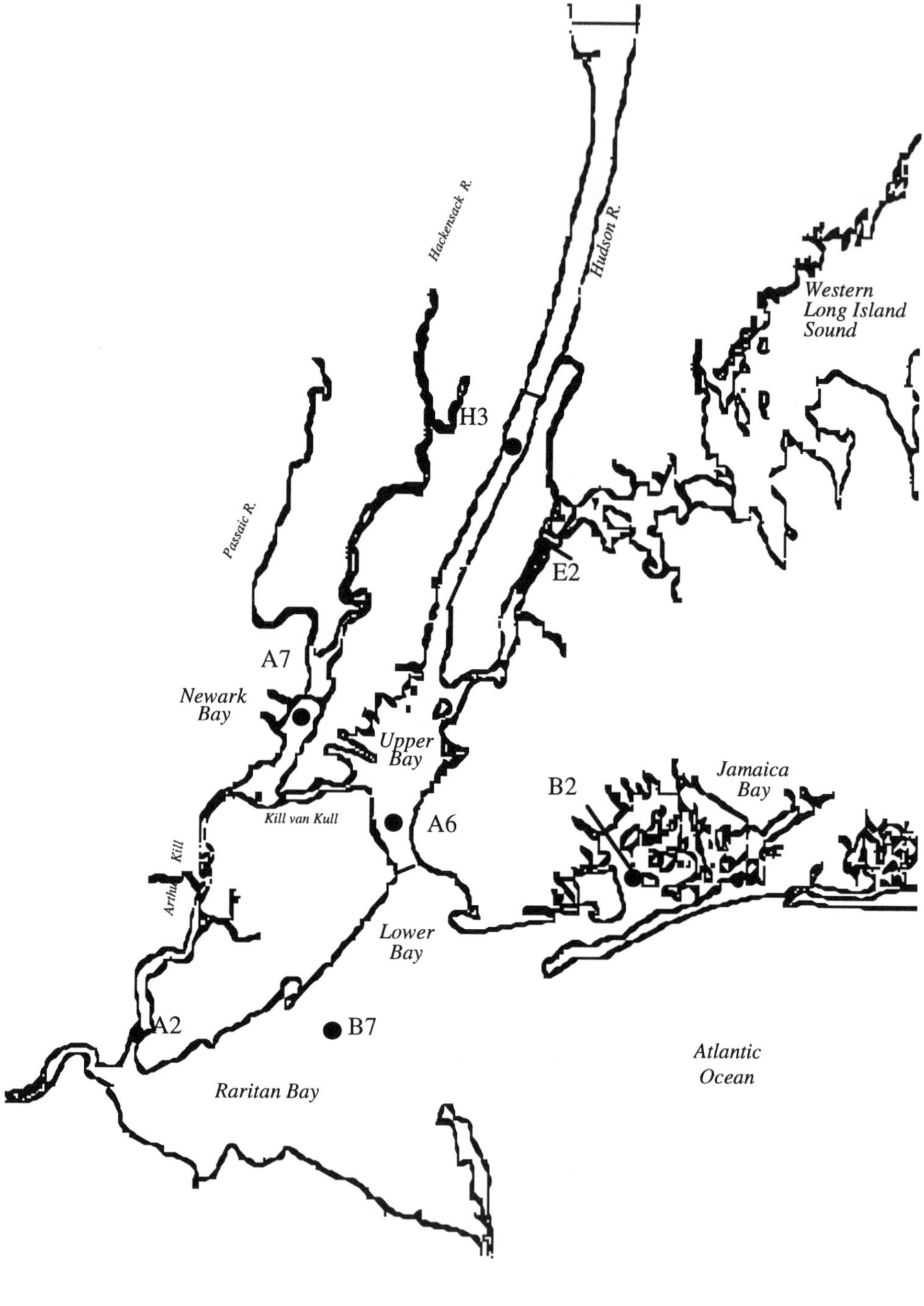

Figure 2. ***Hudson/Raritan Estuary. Sampling stations for copper site-specific study.***

5. SPECIES

- *Mytilus edulis* (blue mussel) - 48-hr embryo-larval development test;
- *Mulinia lateralis* (coot clam) - 48-hr embryo-larval development test;
- *Arbacia punctulata* (sea urchin) - 48-hr embryo-larval development test;
- *Mysidopsis bahia* (mysid shrimp) - 96-hr acute survival test; and
- *Champia parvula* (red algae) - sexual reproduction test.

5.1 Results of the Study

The October 1992 survey/toxicity testing event was fully successful, except that a spawning population of *Mytilus edulis* was not available. Based on the results of the October 1992 event, the April/May 1993 survey/toxicity testing event was modified by eliminating the insensitive *Mysidopsis bahia.* Poor reproductive success in the April 1993 controls for both *Mytilus edulis* and *Champia parvula* resulted in invalid tests. It, therefore, became necessary to retest *Mytilus edulis* and *Champia parvula* in May 1993. The May 1993 *Mytilus edulis* testing was fully successful but *Champia parvula* continued to demonstrate poor reproductive success in the controls and was not retested again [9].

Statistical analyses were conducted at the EPA's ERL-N [10]. Comparisons between laboratory and harbor site water revealed significant differences supporting (control) the conclusion that a site-specific copper criteria could be developed for the Harbor. Species-specific comparisons of toxicity data among the seven Harbor sites indicated that the endpoints were similar and within expected experimental error of toxicity testing and, therefore, supported the development of a single WER for the entire Harbor.

The species-specific WERs are the geometric mean of the 9 LC50s calculated for each species throughout the harbor (6 sites plus 3 tidal samples at the Hudson River site = 9 harbor toxicity tests per species per event). For *Mulinia lateris*, the one species tested during both major sampling events, the WER is the geometric mean of the 18 pooled endpoints from both sampling events. Where there was more than one Narragansett Bay control test conducted on a species during one testing event, the geometric mean of these values was used in the WER calculation.

The species used to develop a site-specific criteria should have sensitivities near but not below the criterion (Figure 3). Therefore, it was decided that *Champia parvula*, a non-commercially important algae with a sensitivity to copper below the criterion and *Mysidopsis bahia*, with a sensitivity well above the criterion, would not be used to develop the final WER. The final WER developed was the geometric mean of *Mytilus edulis* (1.54), *Mulinia lateralis* (1.42) and *Arbacia punctulata* (1.45). The final WER is 1.5 (Table 1).

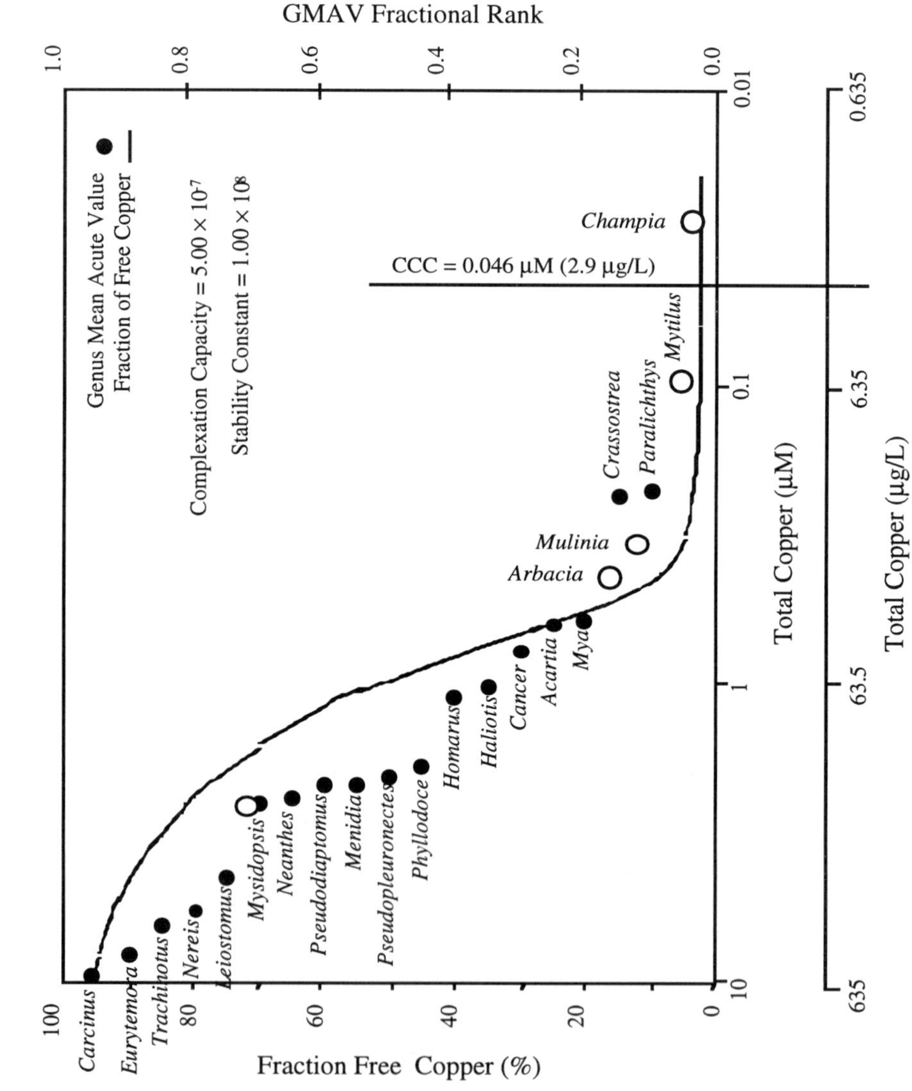

Figure 3. Species Used to Develop a Site-Specific Criteria.

Table 1. Summary of toxicity testing data.

SPECIES	Event[1] Test Results			Geometric Mean IC/LC50s (µg/L Dissolved)		Geometric Mean Water Effect Ratio (WER)	WERs Used to Calculate the Final WER
	First	Second	Third	Narragansett	NY Harbor		
Champia parvula[2]	yes	fail	fail	3.0	6.5	2.17	-
Mysidopsis bahia[2]	yes	no	no	164.0	218.0	1.33	-
Arbacia punctulata	yes	no	no	21.4	31.0	1.45	1.45
Mulinia lateralis	yes	yes	no	17.7	25.2	1.42	1.42
Mytilus edulis	no	fail	yes	12.4	19.1	1.54	1.54
						FINAL WER	1.5

[1]yes = adequate test; no = no test conducted; and fail = test conducted but failed due to poor reproduction in controls.

[2]*Champia parvula* and *Mysidopsis bahia* results were not used to develop the final WER.

6. NATIONAL COPPER CRITERIA

The present national acute copper criterion of 2.9 µg/L total recoverable copper is based on one nominal (unmeasured) *Mytilus edulis* toxicity test endpoint of 5.8 µg/L total recoverable [11]. EPA guidelines for the development of numeric criteria [12] state that in cases where the final acute value (FAV) is greater than the genus or species mean acute value (GMAV or SMAV) of one of the four most sensitive species and the species is an important commercial or recreational species, the GMAV or the SMAV is used rather than the FAV to calculate the criteria.

For copper, the calculated FAV of 5.832 µg/L total recoverable is slightly greater than the one nominal based *Mytilus edulis* SMAV of 5.8 µg/L total recoverable. *Mytilus edulis* is an important commercial and recreational species. Therefore, the *Mytilus edulis* species mean acute value (SMAV) of 5.8 µg/L total recoverable was used to calculate the marine acute copper criteria as follows:

Acute Criterion = SMAV *Mytilus edulis*/2

= 5.8 μg/L total recoverable/2 = 2.9 μg/L total recoverable

The marine chronic criterion is the product of the FAV divided by the acute to chronic ratio (ACR). The copper criteria document [11] did not include a calculated ACR for marine copper. Instead it used the "Factor of 2" assumption. The "Factor of 2" assumption was based on a number of ACRs developed using fish embryo/larval acute and chronic tests. The data from fish embryo/larval tests indicated that an ACR of about 2 could be expected. Since sufficient acute/chronic test data on sensitive marine species had not been conducted and the copper FAV was based on an embryo/larval *Mytilus edulis* test, it was assumed that the "Factor of 2" would be protective of chronic impacts due to copper [11]. Therefore, the chronic marine copper criterion was developed as follows:

Chronic Criterion = SMAV *Mytilus edulis*/ACR

= 5.8 μg/L total recoverable/2 = 2.9 μg/L total recoverable

6.1 Recalculation of the National Criteria

After review of the Harbor toxicity testing data and the data used to develop the national copper criteria, a decision was made to recalculate the national marine copper criteria. It was agreed that the recalculation would be conducted according to the EPA's "Guidelines for Deriving Numerical National Water Quality Criteria for the Protection of Aquatic Organisms and Their Uses." [12].

A literature search was conducted and a request was sent to all coastal EPA Regions for any available marine copper specific toxicity testing data. The majority of the data obtained through the literature search did not comply with the requirements outlined in the above guidance [12]. However, 7 new genera or species, which met the requirements of the guidelines [12], were added to Tables 1 and 3 of the copper criteria document [11]. None of these 7 species replaced any of the 4 most sensitive species (Table 2).

All the total recoverable based GMAVs and SMAVs contained in Tables 1 and 3 of the copper criteria document [11] were converted to dissolved values by multiplying the total recoverable values by the dissolved to total recoverable ratio of 0.83 derived from the Harbor control water (Narragansett Bay) toxicity tests.

Table 2. Genera/species added to Tables 1 and 3 of the National Copper Criteria Document [11].

Genera/Species	GMAV/SMAV (dissolved)
Fundulus heteroclitus (mummichog) [13,14]	1,837 µg/L
Cyprinodon variegatus (sheepshead minnow) [15]	368 µg/L
Atherinops affinis (topsmelt) [16]	243 µg/L
Tigriopus californica (copepod) [17]	214 µg/L
Ophryotrocha labronica (polychaete worm) [18]	30,000 µg/L
Neries virens (polychaete worm) [19]	> 249 µg/L
Menidia beryllina (inland silverside) [20]	130 µg/L

EPA Region IX provided recent toxicity testing data for *Mytilus edulis* and *Crassostrea gigas* (Pacific oyster). *Mytilus edulis* and *Crassostrea gigas* are 2 of the 4 most sensitive species identified in the copper criteria document [11]. These new data were obtained from a recent site-specific copper study conducted for San Francisco Bay through the San Francisco Bay Regional Water Quality Control Board [20].

The control test data obtained from the Harbor and San Francisco Bay studies were used to recalculate SMAVs for *Mytilus edulis* and *Crassostrea gigas* by taking the geometric mean of all values obtained in both studies. Control test data from the Harbor study were used to develop a SMAV for *Mulinia lateris*.

The FAV for copper was recalculated, as described in the EPA Guidelines [12], using the data contained in Tables 1 and 3 and the recalculated GMAVs and SMAVs for the above four most sensitive species. The recalculated FAV was 10.58 µg/L dissolved. The Workgroup decided to use the FAV of 10.58 µg/L dissolved rather than the *Mytilus edulis* SMAV of 9.625 µg/L dissolved because all of the Harbor toxicity endpoints for *Mytilus edulis* ranged from 16.6 to 23.6 µg/L dissolved, which are all above the recalculated SMAV (Harbor and San Francisco Bay).

Table 3. Comparison of the updated GMAVs with the 1984 Copper Criteria Document for the four most sensitive genera.

Species	New York/ New Jersey Harbor SMAVs (μg/L dissolved)	San Francisco Bay SMAVs (μg/L dissolved)	1984 Copper Criteria Document GMAVs (μg/L total recoverable)	Recalculated SMAVs/GMAVs for the Most Sensitive Species (μg/L dissolved)
Mytilus edulis (blue mussel)	12.4[1]	7.24[1]	5.8[2]	9.625[3]
Paralichthys dentatus (summer flounder)	-	-	13.93[1]	1.56[4]
Mulinia lateris (coot clam)	17.10[1]	-	-	17.70[1]
Crassostrea	-	-	14.92[1]	1.38[5]
Crassostrea gigas (Pacific oyster)	-	17.80[1]	7.807[1]	
Crassostrea virginica (Eastern oyster)	-	-	28.52[1]	

[1]Geometric mean of a number of toxicity testing endpoints.

[2]Nominal endpoint from a single toxicity test.

[3]Geometric mean of the NY Harbor and San Francisco Bay toxicity testing endpoints excluding the 5.8 μg/L nominal endpoint from the EPA copper criteria document.

[4]SMAV from the EPA copper criteria document adjusted from total recoverable to dissolved (13.93 total recoverable × 0.83 = 11.56 dissolved).

[5]GMAV of *Crassostrea, Crassostrea gigas,* and *Crassostrea virginica* adjusted as necessary to dissolved.

The Workgroup, based upon more current data [21], decided that the "Factor of 2" ACR used in the EPA copper criteria document [11] was no longer a valid assumption. Since no new chronic data were available, the Workgroup decided to use the freshwater ACR of 2.823.

7. FINAL NEW YORK-NEW JERSEY SITE-SPECIFIC MARINE COPPER CRITERIA

The final New York-New Jersey Site-Specific Marine Copper Criteria were developed as follows:

Harbor Site-Specific Marine Acute Copper Criterion =
= FAV / 2 × WER
= 10.58 µg/L dissolved / 2 × 1.5 = 7.9 µg/L dissolved

Harbor Site-Specific Marine Chronic Copper Criterion =
= FAV / ACR × WER
= 10.58 µg/L dissolved / 2.823 × 1.5 = 5.60 µg/L dissolved

The above criteria are specific to New York-New Jersey Harbor and are not applicable to any other surface water. However, with just a few adjustments, revised national marine acute and chronic criteria, which may be applied to any marine surface water, may be developed.

8. PRESENT STATUS OF SITE-SPECIFIC CRITERIA

The states of New York and New Jersey are currently in the process of adopting the above site-specific copper criteria for the waters of New York-New Jersey Harbor. TMDLs/WLAs/LAs have been developed based on the site-specific copper criteria and will be implemented once the criteria have been adopted by the states.

Acknowledgments: The New York/New Jersey Harbor (Harbor) Site-Specific Marine Copper Criteria Study (the Study) was co-funded by the U. S. Environmental Protection Agency (EPA) and the New York City Department of Environmental Protection (NYCDEP). The New York State Department of Environmental Conservation (NYSDEC), the New Jersey Department of Environmental Protection (NJDEP), the Interstate Sanitation Commission (ISC), the Citizens and the Science and Technical Advisory Committees (CAC and STAC) of the New York/New Jersey Harbor Estuary Program (HEP), the many representatives of the local municipal and industrial dischargers, and the EPA's Environmental Research Laboratory-Narragansett participated and contributed to the development of the Harbor Site-Specific Marine Copper Criteria.

REFERENCES

1. U.S. EPA, "Guidelines for Deriving Numerical Aquatic Site-Specific Water Quality Criteria by Modifying National Criteria," EPA-600/3-84-099 NTIS (Springfield, VA 1984).
2. Squibb, K.S., J.M. O'Connor, and T.J. Kneip, "Toxics Characterization Report," Prepared for Module 3.1 of the New York/New Jersey Harbor Estuary Program. U.S. EPA Region II (New York, NY 1991).
3. Battelle Ocean Sciences, "Evaluation of Trace-metal Levels in Ambient Waters and Discharges to New York/New Jersey Harbor for Waste Load Allocation," Contract# 68-C8-0105, Work Assignment# 2-202, U.S. EPA Region II (New York, NY 1991).
4. Battelle Ocean Sciences, "Evaluation of Trace-metal Levels in Ambient Waters and Tributaries to New York/New Jersey Harbor for Waste Load Allocation," Contract# 68-C8-0105, Work Assignment# 3-213, U.S. EPA, Region II (New York, NY 1992).
5. Battelle Ocean Sciences, "Evaluation of Trace-Metal Levels in Ambient Waters and Tributaries to New York/New Jersey Harbor for Waste Load Allocation," New York City DEP (New York, NY 1992).
6. U. S. EPA, "Development of Site-Specific Copper Criteria for the NY/NJ Harbor Complex Using the Indicator Species Procedure," U.S. EPA, Region II, Support Document, Water Management Division (New York, NY 1994).
7. U. S. EPA, "Total Maximum Daily Load (TMDLs) for Copper, Mercury, Nickel and Lead in NY-NJ Harbor," U.S. EPA, Region II, Support Document, Water Management Division, (New York, 1994).
8. U.S. EPA, *Water Quality Standards Handbook*, Office of Science and Technology, (Washington, DC 1983).
9. Science Applications International Corporation, "Toxicity Testing to Support the New York/New Jersey Harbor Site-Specific Criteria Study," Contract# 68-C8-0006, Work Assignment# C-4-94, U.S. EPA, Region II (New York, NY 1993).

10. Hansen, D.J., "Results of Statistical Analyses Conducted on IC50s and WERs," Personal Communication (1993).

11. U.S. EPA, "Ambient Water Quality Criteria for Copper," EPA-440/5-84-031 NTIS (Springfield, VA 1984).

12. U.S. EPA, "Guidelines for Deriving Numerical National Water Quality Criteria for the Protection of Aquatic Organisms and Their Uses," NTIS# PB85-227049 (Springfield, VA 1985).

13. Gardner, G.R. and G. Laroche, "Copper Induced Liasons in Estuarine Teleosts," *J. Fish. Res. Board Can.* 30:363-368 (1973).

14. Dorfman, D., "Tolerance of *Fundulus heteroclitus* to Different Metals in Salt Waters," *Bull. N.J. Acad. Sci.* 22:21-33 (1977).

15. Hughes, M.M., M.A. Heber, G.E. Morrison, S.C. Schimmel and W.J. Berry, "An Evaluation of a Short-term Chronic Effluent Toxicity Test Using Sheepshead Minnow (*Cyprinodon variegatus*) Larvae," *Environ. Pollut.* 60:1-14 (1989).

16. Anderson, B.S., D.P. Middaugh, J.W. Hunt, and S.L. Turpen, "Copper Toxicity to Sperm, Embryos, and Larvae of Topsmelt (*Atherinops affinis*), with Notes on Inducing Spawning," *Mar. Environ. Res.* 31:17-35 (1991).

17. O'Brien, P., H. Feldman, E.V. Grill and A.G. Lewis, "Copper Tolerance of the Life History Stages of the Slashpool Copepod *Tigriopsis californicus* (Copepoda, Harpacticoida)," *Mar. Ecol. Prog. Ser.* 44:59-64 (1988).

18. Brown, B. and M. Ahsanullah, "Effect of Heavy Metals on Mortality and Growth," *Mar. Pollut. Bull.* 2:182-187 (1971).

19. Raymount, J.E.G. and J.S. Shields, "Toxicity of Copper and Chromium in the Marine Environment," *Int. J. Air Water Pollut.* 7:435-443 (1963).

20. ToxScan, Inc., "Results of Provision E5F Spiked Metals Toxicity Testing," February 14-21, (1991).

21. Hansen, D.J., EPA Memorandum to Richard Caspe, U.S. EPA Region II, April 15 (1994).

ACCURACY IN ANALYSIS - IMPORTANCE OF CLEAN METAL SAMPLING AND ANALYSIS

Carlton D. Hunt
Battelle Ocean Sciences
Duxbury, MA 02332

1. INTRODUCTION

In the early 1970s chemical oceanographers recognized that extraordinary efforts were necessary to control extraneous contamination when conducting metals measurements in sea water [1-3]. Prior to the development of the sampling and analytical techniques that are now commonly referred to as "clean" metals methods, oceanographic data sets were characterized by high variability and poor correspondence with data representative of known biological and physical processes. By the late 1970s improvements in analytical techniques and instrumentation, coupled with recognition and control of sample contamination during collection, storage, and processing, enabled marine scientists to develop data sets that were consistent with known biological and physical processes operating in the oceans and estuaries [4]. As a result, major advances in our knowledge of oceanic and estuarine metals cycles in the oceans, transport processes, geochemical processes, and fate were made. The ability to accurately measure metals at low concentrations also enabled oceanographers to develop an understanding of man's influence on metals distributions in estuaries and oceans, and to better quantify metal sources and inputs to the environment [4].

Chemical oceanographers were forced to develop contamination control strategies because the media they worked with had salt levels (≈3.5%) that significantly interfered with available analytical instrumentation, limiting the ability to detect metals at the concentrations in the low µg/L (ppb) to low ng/L (pptrillion) range. Prior to the advent of the clean methods, the concentration of metals in water and tissue samples analyzed by oceanographers was generally less than that contributed through the acids and reagents used to preserve and process the samples [5]. Even the deionized water used for reagent preparation often had concentrations that exceeded metals levels in many of the samples being analyzed. Contribution of metals from sample collection equipment and storage bottles were also found to be major roadblocks to accurate metals analysis [1,6,7]. Fortunately, proper selection of materials that contact samples, when combined with meticulous procedures for cleaning equipment and storage bottles plus reliable deployment techniques, can control unwanted introduction of metals into samples. These procedures resulted in consistently low and quantifiable procedural blanks, thus reduced the frequency of extraneous, and often irregular, contamination that had caused high variability in early data, and that had made interpretation of early oceanographic metals data difficult.

Other than a few instances where oceanographers crossed disciplines and became involved in non-oceanographic studies [8], the value of and need for clean metals measurement techniques in the analysis of aquatic media did not receive widespread acknowledgement, and was not fully recognized in the regulatory arena until recently [9]. The reasons behind this belated recognition and the slow transfer of the oceanographic clean metals technology to other water media lie in part to disciplinary boundaries (freshwater chemists often don't read the oceanographic literature and vice versa), to training (analytical chemists focus on instrumental technique and interference control rather than on contamination control), to perceptions (wastewater had to have high metals concentrations simply because it was waste), to the relatively high method detection and quantification limits established for methods supporting environmental regulations and permits, or to compliance monitoring that only required quantification of metals concentrations at or above levels established by technology based permits. The perception that most metals data are of generally poor quality [9] has also been amplified because various investigations have attempted to use data intended solely for compliance monitoring to develop loading estimates or to assess trends or to evaluate improvements resulting from management actions. The frustrations experienced by investigators, particularly those trained to measure the lowest possible concentrations, in trying to use data collected for purposes other than intended has frequently been translated to mean the data was bad when, in fact, the data was never intended for rigorous scientific assessment.

Regardless of the past data requirements and uses, present regulations that focus on water quality based effluent limits and watershed planning (Table 1) demand high quality data [9]. High quality data is particularly required to develop fair and equitable total maximum daily loads (TMDLs) and wasteload allocations (WLA) that ensure water quality standards are met. The EPA has determined that the data required to achieve these ends must be acquired using methods that can detect metals at levels well below applicable water quality criteria (Table 2). This further compels that data collected in support of these regulations be accurate and of high precision. Without such data, it is impossible to adequately evaluate source strengths, to correctly determine the status of the receiving waters, to properly determine the fate of metals in the environment, or to develop cause/effect linkages.

Table 1. Representative regulatory and scientific applications requiring analysis using clean metals techniques.

Water Quality Management

1. Total maximum daily load determinations(TMDLs)/Waste Load Allocations (WLA)
2. Compliance with NPDES water quality based permit limits
3. Verification of metals concentrations in effluents
4. Verify/refute metals concentrations in receiving waters.

Site-Specific Aquatic Life Criteria

1. Water effects ratio development
2. Chemical translators (conversion factors between receiving water dissolved and effluent total metals measurements)
   a) Total/dissolved ratios
   b) Partition coefficients (dissolved and particulate)
   c) Geochemical Speciation

Bioaccumulation Studies

1. Laboratory
2. Field

Metal Speciation Studies/Toxicological Testing

Fate and Transport Studies

Table 2. Method detection limits (MDL) and minimum levels (ML) required for analysis of waters under recently issued EPA Clean Metals Methods [10]. Water Quality Criteria are the lowest applicable to water and waste water.

Method	Metal	MDL (µg/L)	ML (µg/L)	Lowest EPA Water Quality Criteria (µg/L)
1631 (Oxidative/Purge & Trap/ CVAFS)	Mercury	0.00005	0.0002	0.012
1632 (Hydride AA)	Arsenic	0.002	0.005	0.018
1636 (Ion Chromatography)	Hexavalent Chromium	0.23	0.5	10.5
1637 (CC/STGFAA)	Cadmium	0.0074	0.02	0.32
	Lead	0.036	0.1	0.14
1638 (ICP/MS)	Antimony	0.0097	0.02	14
	Cadmium	0.013	0.1	0.32
	Copper	0.087	0.2	2.5
	Lead	0.015	0.05	0.14
	Nickel	0.33	1	7.1
	Selenium	1.2	5	5
	Silver	0.029	0.1	0.31
	Thallium	0.0079	0.02	1.7
	Zinc	0.14	0.5	28
1639 (STGFAA)	Antimony	1.9	5	14
	Cadmium	0.023	0.05	0.32
	Trivalent Chromium	0.1	0.2	57
	Nickel	0.65	2	7.1
	Selenium	0.83	2	5
	Zinc	0.14	0.05	28
1640 (CC/ICP/MS)	Cadmium	0.0024	0.01	0.32
	Copper	0.024	0.1	2.5
	Lead	0.0081	0.02	0.14
	Nickel	0.029	0.1	7.1

Moreover, because the EPA now recommends and many states apply water quality criteria and standards to the dissolved phase [9], while National Pollution Discharge Elimination System (NPDES) effluent limits are established on the basis of total recoverable metals measurements, a translator between the dissolved and total phases is necessary to establish effluent limits [11]. Several techniques are allowed under the chemical translator approach, all of which require accurate estimates of metal concentrations for successful application.

For example, the speciation model and partitioning coefficient approaches can only give accurate results if the concentrations used in the models are accurate and reflect true concentrations in the receiving waters. Similarly, direct measurement of the dissolved and particulate (or total) concentrations to estimate the metal partitioning in receiving waters demands that high quality data be available [12].

Finally, within the basic sciences, continued advances in our understanding of metal availability to organisms, and ultimately risk assessments, compel acquisition of very high quality data. More importantly, application of metals speciation to estimates of bioavailiblity is now receiving more attention [13]. Validation of metal speciation and risk models will require measurement of chemical species, placing increased burden on analytical chemists performing these low level measurements.

This paper provides a brief review of the history of the development of clean metals sampling and analysis methods, and provides through selected examples, evidence that properly collected and analyzed samples can result in better information on effluent quality, and better understanding of the concentrations and behavior of metals in coastal systems. These data are provided to support the premise that routine use of clean metals methods is necessary for an adequate scientific basis for water quality regulations, assessments, and enforcement actions and for scientific investigation.

2. HISTORY

Although chemical oceanographic researchers developed clean metals methods almost two decades ago, few non-oceanographers, particularly those conducting monitoring, had knowledge of the techniques and had not applied the methods to routine monitoring (Table 3). This began to change in the early 1990s when the quality of metals data collected under required NPDES and under national river and stream monitoring efforts came into question after New York/New Jersey Harbor was declared impaired under Section 304(l) of the Clean Water Act [14]. This declaration was based on copper and mercury concentrations in the Harbor waters that reportedly exceeded water quality criteria and proposed state standards, even though substantial discrepancies in the metals concentrations reported by a number of studies were evident [15], and preliminary modeling to establish total maximum daily loads (TMDL) and wasteload allocations (WLA) using available point and nonpoint source data could not be reconciled against the measured data. This prompted EPA Region II to conduct a sample collection and analyses effort employing the trace metal clean methods routinely used by the oceanographic community [14,16]. These efforts were intended to provide highest quality data and to provide a better sense of which data was most appropriate for the modeling. The results from this study [16-20] demonstrated that much of the historical monitoring data in the Harbor were suspect and that most metals concentrations measured using clean methods were two to over ten times lower than those measured without the benefit of clean techniques. Although investigations conducted during this study suggested that many of the high historic concentrations resulted from sampling artifacts, the data obtained from intercomparison exercises indicated that they were more likely caused from uncontrolled analytical interferences at the instrument level [21]. At about the same time, Windom *et al.* [22] published independent data for the Hudson River suggesting that metals concentrations developed under the USGS long-term river monitoring were much higher than those measured using clean methodologies. Direct comparison between metals data collected using historical USGS methods and clean techniques demonstrated that the historical methods gave clearly higher results [23]. Since this time, the USGS has also developed and published detailed clean metals sampling and analysis protocols that effectively control extraneous contamination from entering samples [24].

Because of the uncertainties in data quality in the receiving waters, reported concentrations in the point sources to NY/NJ Harbor were also suspect. Thus, the EPA also required sampling and analysis of sewage effluents (using trace metal clean methods) that were discharged to this water body to confirm that reported NPDES monitoring results were in fact accurate and appropriate for estimating metals inputs to the Harbor. These comparisons showed that the detection limits for several metals, especially mercury, were too high to provide data useful to the TMDL determination and that control for contamination artifacts and analytical interference on the effluent samples was often inadequate [21]. As a result of this study, the New York City Department of Environmental Protection (NYCDEP) implemented more stringent sampling and analytical requirements in their sewage treatment facilities and laboratories. Implementation of these protocols has clearly contributed lower measured metals concentrations in the effluents discharged from New York City sewage treatment facilities [17].

Table 3. Synopsis of the historical development of trace metals clean methods to water quality measurements.

- Pre-1970: No attention to "clean methods"; accuracy assumed.
- 1970: C. C. Patterson demonstrates severe contamination in lead analysis; questions oceanographers' ability to accurately analyze metals in oceanic waters.
- 1971 - 1972: NSF Lead intercalibration studies in ocean waters demonstrate that (1) laboratories could not accurately measure lead in ocean waters and tissue routinely and (2) techniques could be modified to improve data quality [2,3].
- 1972 - late 1970s: Oceanographers refine sampling, processing and analysis protocols to ensure oceanographically consistent results [4].
- Late 1970s: University marine laboratories validate acceptable methods for accurate and consistent results in the low µg/L to ng/L range.
- 1980s: Application of clean methods and consistency among laboratories and known oceanic processes become the standard for oceanographers conducting trace metals research.
- Late 1980s: Transfer of clean methods to some fresh water researchers, especially for Hg [8]; EPA methods acknowledge need for clean methods but limited guidance and transfer to agencies and dischargers conducting routine monitoring.
- Early 1990s: Demonstration that clean methods were required in the regulatory arena and for routine river, lake, and ocean receiving water monitoring as well as water quality base environmental management [16,17,22].
- 1993: EPA recognizes (1) Water quality based environmental management requires clean methods; (2) Dissolved phase more likely represents bioavailable metals forms than total measurements [9].
- 1993 - 1995: Publication of studies demonstrating the value of the metals contamination control in regulatory, monitoring, and research arena [12,23,25,26].
- 1994: EPA issues guidance for developing site-specific water quality criteria including use of "clean methods" for chemical measurements during toxicological testing [27].
- 1994: EPA develops detailed guidance for measuring metals at concentrations at least 5 times the lowest applicable aquatic effects criteria.
- 1995: EPA Office of Science and Technology publishes 1600 Series Analytical Methods for water and waste water analysis using "clean methods" (See Table 2).

As a result of this and other evidence that application of clean techniques results in lower reported concentrations, demand for better guidance on the clean techniques increased. Concern over the ability of analytical laboratories to conduct measurements correctly persuaded the EPA and others to provide more guidance on the metal clean sampling and analysis methods than previously available [9,27]. More specific guidance on how to conduct clean metals sampling, processing, and analysis can be found in recently published EPA guidance documents that detail the requirements and methods for clean sampling [10,28,29] and for achieving detection method limits at least 5× below the lowest available Water Quality Criteria (Table 2). The interested reader is referred to these methods and many of the references included in this paper for specifications and requirements for the clean metals methods.

3. IMPORTANCE OF QUALITY DATA

Many questions continue to be raised relative to applicability and need for clean metals methods. The question most often asked from those responding to the regulatory demands for higher quality data is "When do I need to use clean metals methods, and why?" The answers to these questions depend on several factors including planned data usage, quality of available data, expected metals concentrations, the specific metal being measured, and the concentrations in the water being measured. In its most stringent application, clean metals methods target zero contribution of metals to the sample during collection, processing, and analysis. This goal is most relevant to metals measurements in the very low part per billion to sub part per trillion range. Mercury concentrations in rivers, lakes, and the oceans have consistently been found in this range when clean methods are employed [8], whereas effluent concentrations appear to range from 5 to 200 ng/L [19,20,30-32]. For higher concentrations (low to mid µg/L), clean metal methods should strive to control extraneous contamination to a known, consistent contribution to the sample that never exceeds 10% of the lowest concentrations measured.

Ultimately, the decision to use clean metals methods and the level of clean technique applied depends on specific regulatory requirements, the environmental management goals being addressed, and the availability of previous known high quality data. For some routine measurements, simply developing and validating the historical data against clean methods results may be all that is required. For others, an acute awareness of potential sources of contamination to samples during collection and processing has been shown to significantly reduce the analytical blank to acceptable levels and may be all that is required to achieve better analytical results. For others, development of a rigorous clean metals analytical approach will be required. Regardless, the principles of clean methods outlined in the above references should be the normal operating philosophy of all analytical laboratories performing low level metal sampling and analysis.

Three major data uses are relevant to the questions posed above. They include 1) compliance monitoring, 2) setting effluent limits, and 3) scientific studies (i.e., chemical speciation studies, loading estimates, impact and risk assessment, etc). Application of clean metals methods to these data uses depends in large measure on the risk one is willing to take relative to regulatory action based on poor data quality. This is a clear policy decision for those who are in the regulated community and should be a major consideration when determining the level of clean method to apply. For those working in the basic sciences, the risk from not using clean methods is rejection of data under peer review and scientific credibility. Thus, for researchers studying fundamental chemical, geochemical, and biological process in natural waters, application of clean methods is imperative and inflexible.

The why question posed above is next addressed through a series of examples that show the value of the clean methods and resulting improvements in data quality and comparability. These examples focus on compliance and scientific data usage. The value and importance of clean methods for setting effluent limits and addressing permit conditions are evident in a number of recent papers [17,25,26] and are not discussed further in this paper. Compliance issues and scientific considerations are discussed next.

Compliance monitoring: Measurements conducted in support of NPDES permit monitoring generally require detection of metals at levels that are high relative to the measurement capabilities of modern instruments or concentrations in receiving waters. For example, Table 4 shows the method detection limits (MDLs) in the NPDES permit issued to the Massachusetts Water Resources Authority for its primary treated effluent in 1993 [31]. Results from the NPDES monitoring of total recoverable metals and for sample splits concurrently collected, but processed and analyzed using clean metals procedures, are also provided. Note that this data only provides a comparison between laboratories, as samples were split after collection from a 24-h sample composite. In this comparison, generally good results between the two analytical facilities were obtained for metals present in the effluent at concentrations well above the relatively high NPDES reporting levels (i.e., Cu, Zn). Thus, quantification of these metals was relatively straightforward and the analytical methods did not need to be pushed heavily toward the clean techniques to meet the objectives of the intended data use. For other metals, the NPDES detection limits are much higher than the measured concentrations in the effluent, often approaching the detection limits of the graphite furnace atomic absorption analysis used to quantify the metals in these samples. Again the NPDES data set met the intended data use, permit compliance. However, comparability between the analytical laboratories could not be established nor could the NPDES data be used for estimating loading to the receiving waters [33]. Further, if low effluent levels were to be imposed, the NPDES reporting limits would be lower and the level of clean technique required would increase.

Table 4. Comparison of NPDES method reporting limits, method detection limits achieved using clean metals methods, and mean concentrations in the Massachusetts Water Resources Authority Deer Island primary treated effluent [31,32].

	Ag (μg/L)	Cd (μg/L)	Cr (μg/L)	Cu (μg/L)	Pb (μg/L)	Hg (μg/L)	Zn (μg/L)
NPDES monitoring laboratory reporting limit	7	5	10	55	30	0.5	5
Clean metals analysis reporting limit	0.55	0.06	0.2	0.87	1.2	0.025	0.28
Mean measured concentrations in effluent: 1993 - 1994	4.2	0.6	3.4	72	11.8	0.16	87

A second example of the importance of clean metals techniques to environmental measurements is a comparison of copper levels in a waste that was analyzed using routine sample collection methods, but analyzed under both routine and trace metal clean protocols (Figure 1). In this study, routine monitoring results have been normalized to the results from the clean metals procedures, then plotted against the concentrations determined using the clean method. If agreement between the methods is perfect, the resulting ratio would equal one. Any contamination in the routine analysis (high result) is displayed as a positive excursion from unity and any underestimate as a negative excursion. In this example, the agreement between the laboratories performing the analyses is reasonably good at higher concentrations. However, significant departures begin to occur at concentrations in the vicinity of 6 μg/L. Extreme disagreements in concentrations (up to 10 fold) are evident at concentrations near 1 µg/L. In other cases, the agreement between the analytical laboratories breaks down at about 30 µg/L [34]. Similar results, although not shown here have been observed for the metals Cr, Pb, Cd, and Zn [35].

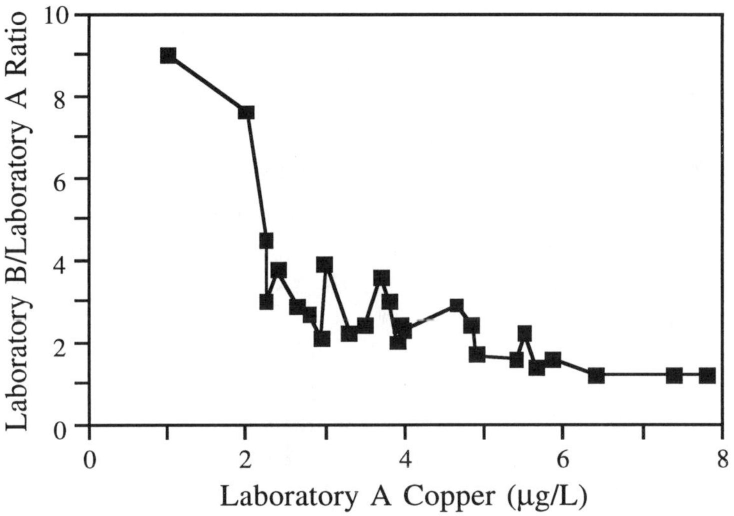

Figure 1. ***Comparison of Copper Results in an Industrial Effluent Using Routine and Clean Metals Methods. A ratio of one (1) indicates perfect agreement; ratios that are >1 indicate high values relative to clean method measurement protocols.***

Scientific investigations: Scientific assessments include development of information necessary (1) to understand the status of a system including chemical and geochemical processes and chemical speciation, (2) to develop chemical translators between dissolved concentrations in the receiving waters and total concentration measurements in effluents, (3) to evaluate trends in the water quality, (4) to measure chemical speciation, (5) to develop transport and fate models (both for calibration and validation), or (6) to evaluate risk. The data requirements for each of these assessments push the analyst to achieve the lowest possible detection limits and to ensure that the data are as accurate as possible.

Comparison of metals distributions in estuaries provides an example of how high quality data can describe basic geochemical phenomena and provides an understanding of how estuaries differ. Estuarine data are useful in this context because the salinity gradient of estuaries is useful for normalizing data, and can thereby be used to examine similarities and differences in contaminant levels, ranges, and geochemical processes against a well defined conservative tracer of freshwater mixing with seawater. However, such comparisons are only useful if results are uncompromised by sample collection and processing errors, and the data encompass the full range of salinities manifest in the system.

Several high quality estuarine metals data sets meeting these requirements have become available over the past decade. These include, but are not limited to Narragansett Bay, RI and the New York/New Jersey Harbor complex on the east coast [19,30,36], the Mississippi River delta in the Gulf of Mexico [16], and San Francisco Bay and Puget Sound on the west coast [37-39].

For the Narragansett Bay Study, samples were collected during surveys conducted during four periods, two in the fall of 1985 and two in the spring of 1986. Samples were collected from 2 to 4 depths throughout the system (Figure 2) and measured for dissolved and particulate metals plus other basic oceanographic parameters (salinity, temperature, depth, nutrients, total suspended solids). In this system, the distribution and behavior of dissolved copper was generally representative of other metals (silver, nickel, lead, chromium, cadmium) measured during this study. Dissolved copper from the four surveys (Figure 3) shows a consistent concentration range (0.2 to 4.5 µg/L) that systematically decreases from the head of the estuary (lowest salinity) into offshore Rhode Island Sound (highest salinities). The general linear trend in the data and negative slope between copper concentration and increasing salinity is indicative of classical conservative mixing between the fresher portions of the system and coastal waters. Note that copper concentrations at the highest salinities were very similar among all four surveys. More variability in the concentrations at lower salinities (<30‰) was observed. This variability is related to changes in the freshwater flow at the head of the Bay, intense local input from a major treatment plant located at the head of the Bay, plus smaller scale regional differences caused by the island and bay segments characteristic of Narragansett Bay [36].

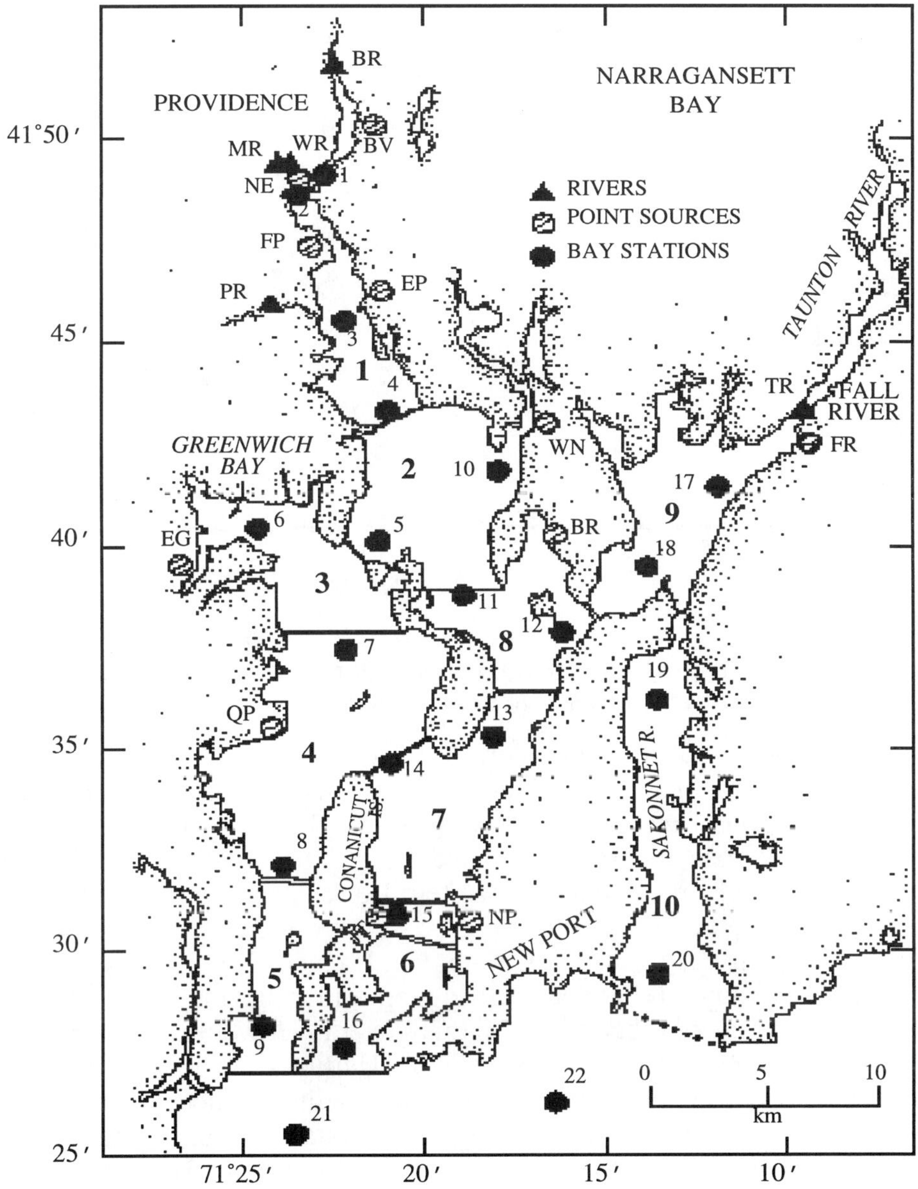

Figure 2. ***Narragansett Bay Water Column Sampling Stations During Bay-Wide Surveys Conducted in the Mid 1980s.***

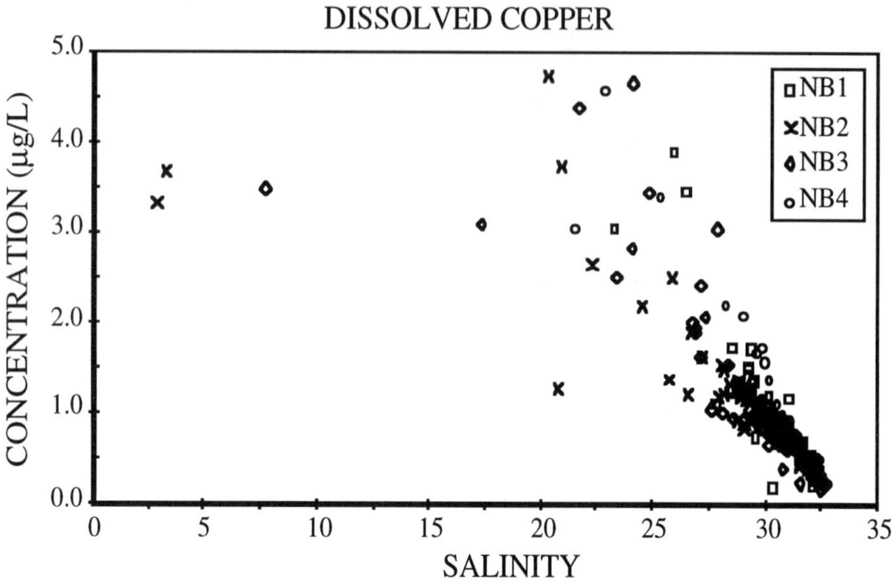

Figure 3. ***Dissolved Copper Versus Salinity from Four Surveys in Narragansett Bay in the Mid 1980s.***

The waters of the New York/New Jersey Harbor complex were sampled during high flow (January 1991) and low flow conditions (October 1991) of the Hudson River [19,30]. Samples from 36 stations distributed across the freshwater reaches of the Hudson River and throughout the Harbor including western Long Island Sound to offshore waters of the New York Bight (Figure 4) were collected and analyzed for eight metals (Ag, As, Cd, Cu, Ni, Hg, Pb, and Zn). As observed in Narragansett Bay, the copper distribution and behavior represents the general response of other metals in this system. However, in contrast to Narragansett Bay, copper in New York Harbor does not show clear evidence of conservative mixing in the mid salinity range (10 to 25%) even though freshwater flow through this system is dominated by the Hudson River (Figure 5). In contrast, the New York Harbor system has maximum copper concentrations between 20 and 25% salinity. Lower copper concentrations were found in the Hudson River above the New York City complex and in the higher salinity offshore regions. This apparent maximum in dissolved copper represents input from numerous sewage treatment plants and other discharges located throughout the harbor complex, and which dominate local freshwater flow and copper loading to the Harbor. These inputs clearly override any conservative mixing that might occur between the Hudson River and offshore waters. This observation is confirmed by the generally higher concentrations measured in this system during a low flow condition of the Hudson River (Figure 5). The data suggest that conservative mixing occurs between salinities 25 to 32%. This salinity range occurs downstream of the local inputs and reflects the region of this system where mixing of water from the New York Bight begins to dominate the behavior of copper in the system. Such conservative behavior in dissolved copper in offshore waters along the axis of the Hudson Shell Valley has been previously reported by Hanson and Quinn [40]. Comparison of the dissolved copper concentrations in the Harbor area in 1991 (1 to 3 µg/L) to those measured in the early 1970s (1 to 7 µg/L) by Klinkhammer and Bender [41] suggests lower concentrations (2-fold) in the more recent sampling. Although this earlier study found a decreasing linear trend between salinity and dissolved copper in the Harbor, sporadic high concentrations between 4 and 7 µg/L were measured. They did not report a maximum in the copper concentration, although a peak in dissolved Ni concentrations at about 2% salinity was evident in their data.

The dissolved copper in these two estuaries and the other three estuaries noted above is contrasted in Figure 6. As in the Narragansett Bay and NY/NJ Harbor studies, sample collection stations in these studies covered the entire system and included the entire salinity range (0 to 32%) characteristic of these systems. The correspondence between dissolved copper concentrations and salinity in the mouth of the Mississippi River is clearly and strongly conservative (linear), and shows limited input from sources other than the river. Dissolved copper in San Francisco Bay shows two signatures. One is dominated by river flow in the northern Bay (sequence of lower concentrations between 5 and 32% salinity); the second is in the south Bay (series of rapidly increasing copper at salinities between 31 and 32%). The major fresh water input to this area is dominated by a single sewage treatment plant [38]. This, coupled with restricted circulation in this part of the Bay, leads to relatively high copper concentrations compared to the northern Bay. Reported dissolved copper concentrations in Puget Sound in the early 1980s were generally similar to those observed in the coastal areas of the other estuaries discussed above.

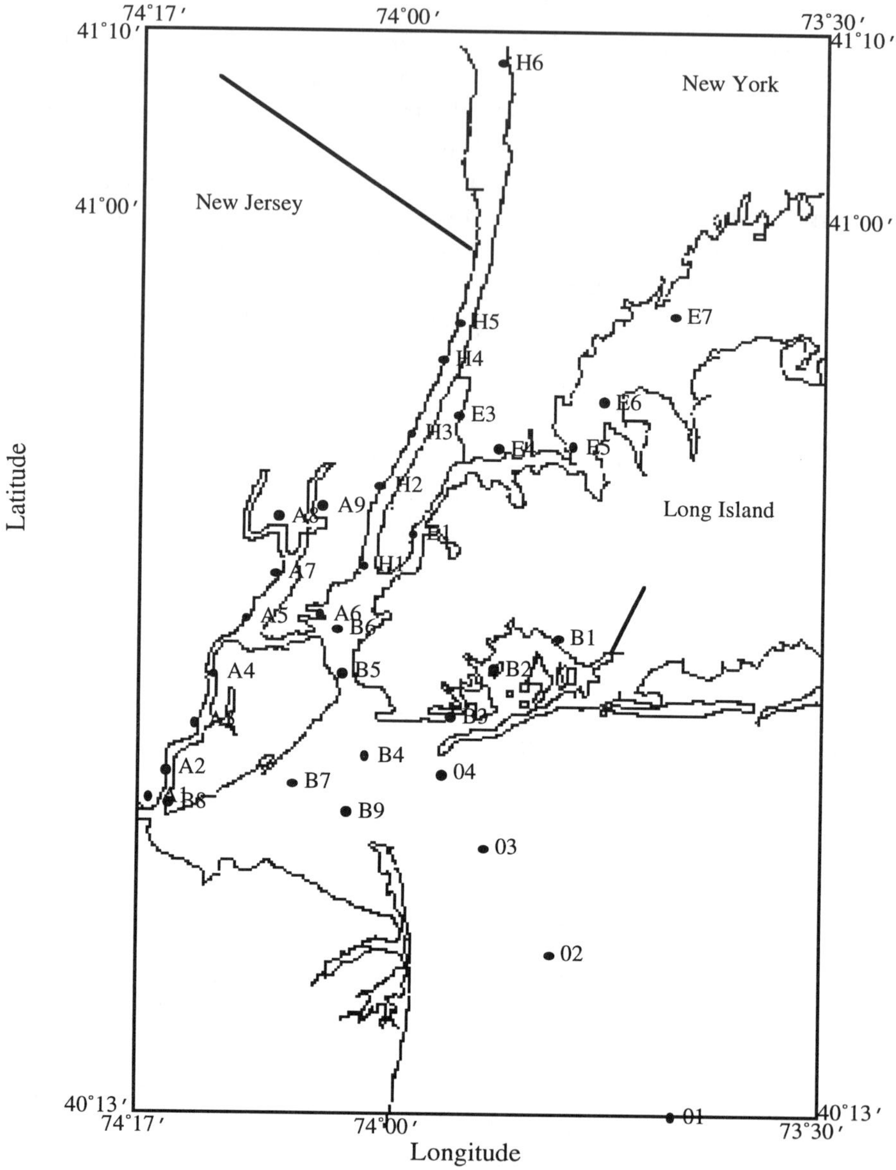

Figure 4. ***New York/New Jersey Harbor Water Column Station Locations, January 1991.***

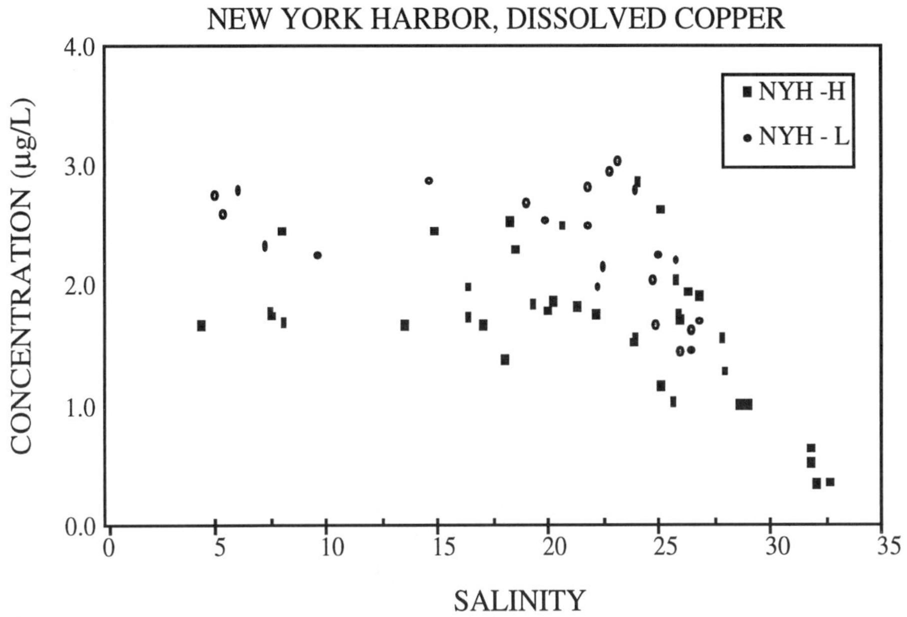

Figure 5. ***Dissolved Copper Versus Salinity from Two Surveys in New York/New Jersey Harbor in 1991.***

Notable in these copper data is the relatively narrow concentration range observed over the entire data set (0.2 to 5 µg/L). This range appcars to rcprcscnt typical copper concentrations in the major estuaries and coastal regions of the United States. Copper concentrations in the offshore coastal waters included in these data sets range from 0.1 to 0.3 µg/L. Copper concentrations in the rivers (0% salinity on the graphs) associated with each of these estuaries are also similar (1 to 2 µg/L). Copper values in the river sources to Narragansett Bay were slightly higher (3 µg/L), but are known to drain a highly industrialized area [42]. Across the five systems represented in Figure 6, higher copper concentrations are distinctly manifest within the systems when treatment facilities dominate the source terms (relative to rivers). It is only near these sources that copper concentrations approach or exceed the marine water quality criteria. Further, in each of these systems, the response of dissolved copper away from the immediate influence of major discharges is reasonably consistent versus salinity and follows predictable geochemical behavior (i.e., conservative mixing).

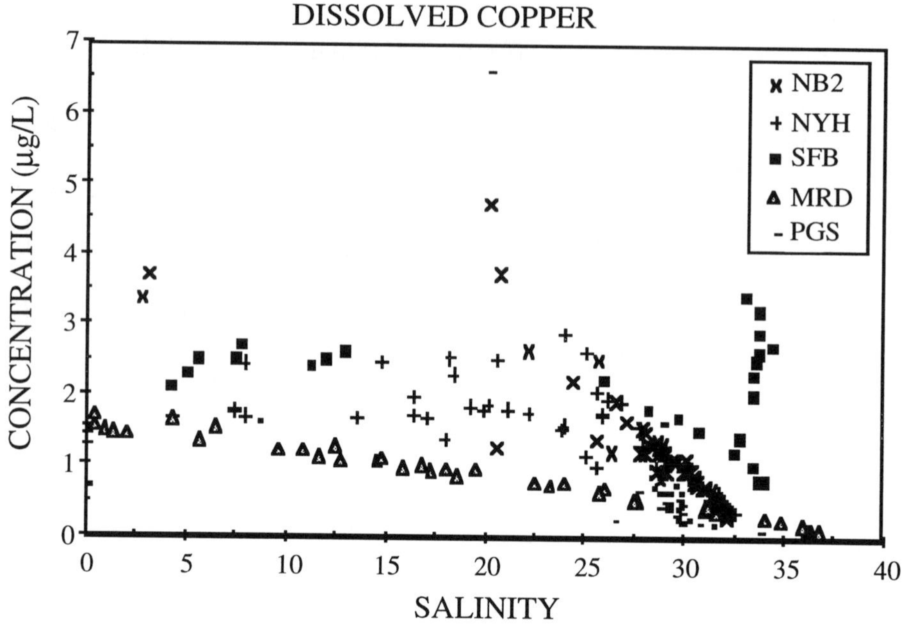

Figure 6. ***Comparison of Dissolved Copper Versus Salinity in Five Major Estuaries of the U.S.***

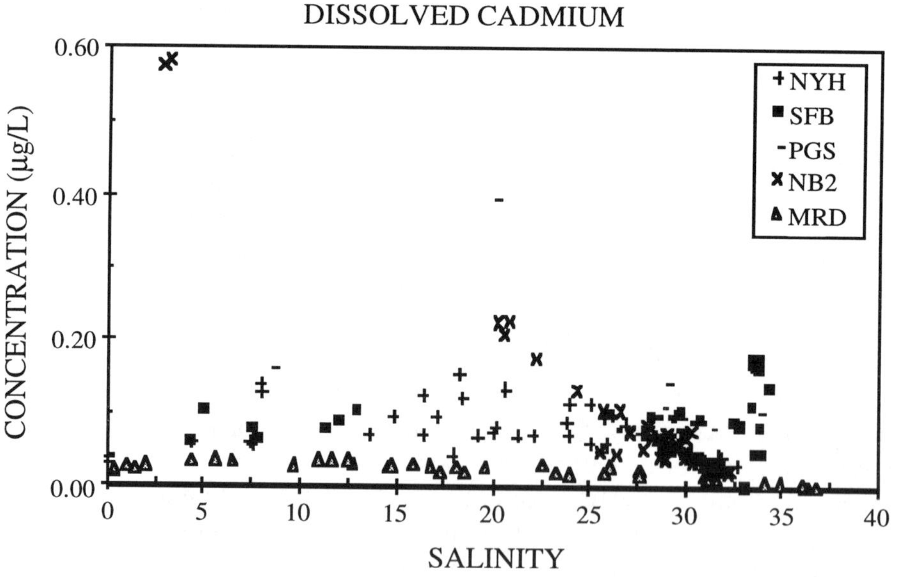

Figure 7. ***Comparison of Dissolved Cadmium Versus Salinity in Five Major Estuaries of the U.S.***

The distribution and response of other metals in these systems follow a similar behavior. For example, Figure 7 compares dissolved cadmium concentrations in these five systems. The cadmium data show generally similar concentrations among the five estuaries (<0.2 μg/L) with some higher values close to major sources, especially in Narragansett Bay. The estuaries in highly industrialized and urbanized regions clearly have higher cadmium concentrations than found in the Mississippi River Delta. As observed for copper, cadmium concentrations in the rivers supplying the freshwater to these systems are similar. This suggests the range of dissolved cadmium in these rivers is not widely divergent. As also seen in the copper data, offshore coastal seawater cadmium concentrations are approximately the same and very low (<0.02 μg/L).

4. CONCLUSION

Clean metal analysis techniques have been available to aquatic scientists for over two decades and as shown above provide scientists with techniques for understanding and comparing metals behavior in diverse systems. However, the application of these techniques to routine measurement of metals in many receiving waters and effluents is a recent phenomenon driven by the need to provide data of known high quality for water quality based permitting decisions and to better understand the fundamental processes controlling the behavior, fate, and bioavailablity of metals. A growing body of data continues to show metal levels in many lakes, rivers, and streams are lower than previously measured. As the levels go down, the need for clean techniques will increase. Thus, those faced with evaluating the quality of receiving waters and with developing effluent limits will need to increase the level of attention paid to extraneous entry of contamination to samples during sample collection, processing, and analysis.

Further, failure to use clean methods when monitoring receiving waters or dealing with potential regulatory action or revising permit limits through published regulatory methods (e.g., chemical translators [11]) can negatively impact study results, and may ultimately result in overly restrictive, and unnecessary effluent limits. Of particular concern is the impact that poor data quality (especially high false positive, but also including high variability) can have on decisions regarding exceedance of state standards and on estimates of the maximum metal loads that can be discharged.

Ultimately, the effect of poor data will be imposition of more stringent source control requirements, which could lead to unnecessary expenditure of capital resources. In addition, availability of high quality data can also assist toxics reduction programs through identification of sources to municipal collection systems and by quantitatively demonstrating that toxic reduction strategies and management actions have resulted in improvements to overall quality of wastes and receiving waters. A case in point is the clearly evident improvements in sewage effluent quality achieved by the New York City DEP through better measurement techniques and demonstrations that toxic reduction strategies are in fact working [17]. Simply put, the use of clean metals methods is no longer only relevant to researchers studying water chemistry or geochemistry. Rather, appropriate use of clean metals methods is now a fundamental requirement for those conducting routine monitoring of effluents and receiving water.

Further, the application of clean methods will benefit environmental regulators and managers by providing more consistent and reliable data from which water quality can be judged and from which management decisions can be made. Better data will also improve transport, fate and effect models, provide better estimates of bioavailability, and result in general improvements in the science underpinning water quality criteria and management actions. Similarly, as risk based environmental management increases, the demand for quality data to drive risk models will also increase, and should result in more reliable and defensible risk assessment. Thus, the need for clean metals methods in water quality based environmental management will increase the demand on those making environmental measurements to use clean methods. These trace metal clean sampling and analytical requirements will receive even greater attention in support of water quality criteria revisions and development and will be of greater concern, if we move to metal speciation based aquatic life criteria.

Acknowledgments: Thanks to Russ Flegal and Allen Shiller for providing their data sets to me. Many students and technical staff from the University of Rhode Island and Battelle Ocean Sciences contributed to the collection and analysis of the data sets from Narragansett Bay and New York Harbor. Their dedication to quality is recognized and gratefully acknowledged.

REFERENCES

1. Patterson, C.C. and D.M. Settle, "The Reduction in Orders of Magnitude Errors in Lead Analysis of Biological Materials and Natural Waters by Evaluating and Controlling Extent and Sources of Industrial Lead Contamination Introduced During Sampling and Analysis," in *Accuracy in Trace Analysis: Sampling, Sample Handling, Analysis*. P.D. LaFleur, Ed., pp. 321-352. National Bureau of Standards Spec. Pub. 422. U.S. Government Printing Office (Washington, DC 1976).

2. Brewer, P., N. Frew, N. Cutshall, J.J. Wagner, R.S. Duce, P.R. Walsh, G.L. Hoffman, J.W.R. Dutton, W.F. Fitzgerald, C.D. Hunt, D.C. Girwin, R.G. Glen, D. Patterson, D. Settle, B. Glover, B.J. Presley, J. Trefry, H. Windom and R. Smith, "Interlaboratory Lead Analyses of Standardized Samples of Sea Water," *Mar. Chem.* 2:69-84 (1974).

3. Bewers, J.M., S. Hartling, R. Duce, P. Walsh, W. Fitzgerald, C.D. Hunt, H. Windom, R. Smith, C.S. Wong, P. Berrang, C. Patterson and D. Settle, "Comparison Determinations of Lead by Investigators Analyzing Individual Samples of Seawater in Both Their Home Laboratory and in an Isotope Dilution Standardization Laboratory," *Mar. Chem.* 4:389-392 (1976).

4. Bruland, K.W., "Trace Elements in Sea-Water," in *Chemical Oceanography*, Vol. 8. pp. 157-220. Academic Press (London, UK 1983).

5. Zief, M. and J.W. Mitchell, "Contamination Control in Trace Element Analysis," in *Chemical Analysis Series*, Volume 47. p. 262. John Wiley & Sons (New York, NY 1976).

6. Bewers, J.M. and H.L. Windom, "Intercomparison of Seawater Sampling Devices for Trace Metals," in *Trace Metals in Sea Water*, C.S. Wong, E. Boyle, K.W. Bruland, J.D. Burton, and E.D. Goldberg, Eds., pp 143-154. Olenum Press (New York, NY 1981).

7. Gill, G.A. and W.F. Fitzgerald, "Mercury Sampling of Open Ocean Waters at the Picomolar Level," *Deep Sea Res.* 32:287-297 (1985).

8. Fitzgerald, W.F. and C.J. Watras, "Mercury in the Surficial Waters of Rural Wisconsin Lakes," *Sci. Total Environ.* 87/88:223-243 (1989).

9. Prothro, M.G., "Policy and Technical Guidance on Interpretation and Implementation of Aquatic Life Metals Criteria," U.S. Environmental Protection Agency, Office of Water, October (Washington, DC 1993).

10. U.S. EPA, "Method 1669: Sampling Ambient Water for Trace Metals at EPA Water Quality Criteria Levels," EPA-821/B-95-034, U.S. Environmental Protection Agency, Office of Water, Engineering and Analysis Division (Washington, DC 1995).

11. U.S. EPA "The Metals Translator: Guidance for Calculating a Total Recoverable Permit Limit from a Dissolved Criterion Manual," EPA-823/B-96-007, U.S. Environmental Protection Agency, Office of Water, Office of Science and Technology (Washington, DC 1996).

12. Leonard, J.N., R.E. Fergen, A.J. Lawler and P.N. Boothe, "Monitoring Trace Metal Concentrations in Virginia Receiving Waters Using the Chemical Translator Approach with Clean Techniques," in *Proceedings of the Water Environment Federation 67th Annual Conference & Exposition*, pp. 509-518. (Chicago, IL 1994).

13. Bergman, H.L. and E.J. Dorward-King, Eds., *Reassessment of Metals Ctiteria for Aquatic Life Protection: Proirities for Research and Implementation* SETAC Press (Pensacola, FL 1997).

14. Locicero, F., W.F. Jackson, P.M. Pergola, T.M. Brosnan, D.J. Hansen, G.B. Thursby, H.E. Allen and D.A. Lewis, "Development of Site-Specific Marine Copper Criteria for New York/New Jersey Harbor," in *Metasl in Surface Waters*, H.E. Allen, A.W. Garrison and G.W. Luther, Eds., pp. 91-105. Ann Arbor Press (Chelsea, MI 1997).

15. Squibb, K., J.M. O'Connor and T.J. Kneip, "New York/New Jersey Harbor Estuary Program Module 3.1: Toxics Characterization," Report to the New York/New Jersey Harbor Estuary Program prepared by the Institute of Environmental Medicine, New York University Medical Center (New York, NY 1991).

16. Hunt, C.D., D.A. West and D.A. Lewis, "Trace Metal Concentrations in New York/New Jersey Harbor," in *Hazardous and Industrial Wastes*, Proceedings of the Twenty-Sixth Mid-Atlantic Industrial Waste Conference, C.P. Huang, Ed. pp. 691-698. Technomic Publishing Co., Inc. (Lancaster, PA 1994).

17. Brosnan, T.M., A.J. Stubin, V. Sapienza and Y.G. Ren, "Recent Changes in Metals Loadings to New York Harbor from New York City Water Pollution Control Plants," in *Hazardous and Industrial Wastes*, Proceedings of the Twenty-Sixth Mid-Atlantic Industrial Waste Conference, C.P. Huang, Ed. pp. 657-666. Technomic Publishing Co., Inc. (Lancaster, PA 1994).

18. Brosnan, T.A. and M.L. O'Shea, *New York Harbor Water Quality Survey: 1991-1992*. City of New York Department of Environmental Protection, NY. 59 pgs. plus appendices (New York, NY 1993).

19. Battelle Ocean Sciences, "Evaluation of Trace-Metal Levels in Ambient Waters and Discharges to New York/New Jersey Harbor for Waste Load Allocation," Final Report, U.S. EPA, Region II (Duxbury, MA 1991).

20. Battelle Ocean Sciences, "Evaluation of Trace-Metal Levels in Ambient Waters and Discharges to New York/New Jersey Harbor for Waste Load Allocation," Final Report, U.S. EPA, Region II (Duxbury, MA 1991).

21. Battelle Ocean Sciences, "Final Results of the Ambient and Municipal Sample Interlaboratory Comparison Study in Ambient Waters and Discharges to New York/New Jersey Harbor," Contract# 68-C8-0105, Work Assignment# 2-202, Report submitted to the U.S. Environmental Protection Agency, Washington, DC and Region II (Duxbury, MA 1991).

22. Windom, H.L., J.T. Byrd, R.G. Smith, Jr. and F. Huan, "Inadequacy of NASQAN Data for Assessing Metal Trends in the Nation's Rivers," *Env. Sci. Technol.* 25:1137-1142 (1992).

23. Taylor, H.E. and A.M. Shiller, "Mississippi River Methods Comparison Study: Implications for Water Quality Monitoring of Dissolved Trace Metals," *Env. Sci. Technol.* 29:1313-1317 (1995).

24. USGS. *A Protocol for the Collection and Processing of Surface-water Samples for Subsequent Determination of Trace Elements, Nutrients, and Major Ions in Filtered Water*, Office of Water Quality Technical Memorandum 94.09, Office of Water Quality, Water Resources Division, U.S. Geological Survey, January (Reston, VA 1994).

25. Waddle, R.D., "Cleaning-Up a Contested Copper Limit - The Solution: Improved Sampling Techniques," *Water Environ. Lab. Sol.* 2(4):4-5 (1995).

26. Waddle, R.D., "Characterization of Trace Metals in Wastewater Effluents and Ambient Receiving Waters Using "Clean" Techniques," in *Environmental Laboratories: Testing the Waters*, pp. 3-13 to 3-24. Water Environment Federation Specialty Conference Series CP3506 (Alexandria, VA 1995).

27. U.S. EPA, "Interim Guidance on Determination and Use of Water Effects Ratios for Metals," U.S. Environmental Protection Agency, Office of Water, Engineering and Analysis Division (Washington, DC 1994).

28. U.S. EPA, "Guidance on Establishing Trace Metal Clean Rooms in Existing Facilities," EPA 821/B-95-001, U.S. Environmental Protection Agency, Office of Water, Engineering and Analysis Division (Washington, DC 1995).

29. U.S. EPA "Guidance on the Documentation and Evaluation of Trace Metals Data Collected for the Clean Water Act Compliance Monitoring," EPA 821/B-95-002, U.S. Environmental Protection Agency, Office of Water, Engineering and Analysis Division (Washington, DC 1995).

30. Battelle Ocean Sciences, "Evaluation of Trace-Metals in New York/New Jersey Harbor Ambient Waters, Tributaries, and Discharges During Low Flow Conditions for Waste Load Allocations," Final Report to New York City DEP, Battelle Ocean Sciences (Duxbury, MA 1992).

31. Uhler, A.D., D. West and C.D. Hunt, "Deer Island Effluent Characterization June - November 1993," Environmental Quality Department, Technical Report Series# 94-4, 46 pgs. Massachusetts Water Resources Authority (Boston, MA 1994).

32. Hunt, C.D., D.E. West and C.A. Peven, "Deer Island Effluent Characterization Studies June 1993 - November 1994," Environmental Quality Department, Technical Report Series# 95-7, 140pgs. Massachusetts Water Resources Authority (Boston, MA 1995).

33. Alber, M. and A.B. Chen, "Sources of Contaminants to Boston Harbor: Revised Loading Estimates," Environmental Quality Department Report, Technical Report Series# 94-1, 93 pgs. Massachusetts Water Resources Authority (Boston, MA 1994).

34. Hunt, C.D. and D.A. Lewis, "To Clean or Not to Clean: Trace Metal Sampling and Analysis and Intercomparisons, Environmental Laboratories: Testing the Waters," Water Environment Federation Specialty Conference Series, Cincinnati OH, August 1995, pp. 3-1 to 3-12. (Alexandria, VA 1995).

35. Hunt, C.D. and D.A. Lewis, "Trace Metal Clean Techniques: Problems, Quality Assessments, and Comparisons," EPA 821/R-95-008, pp. 197-238. Seventeenth Annual EPA Conference on Analysis of Pollutants in the Environment, May 3-5, 1994 (Norfolk, VA 1995).

36. Pilson, M.E.Q. and C.D. Hunt, "A Summary of Results from the SINBADD Cruises 1985-1986," Report# NBP-89-22. 118 pgs. Report submitted to the Narragansett Bay Project, Rhode Island DEM, University of Rhode Island, March (1989).

37. Shiller, A.M., "Comparison of Nutrient and Trace Element Distributions in the Delta and Shelf Outflow Regions of the Mississippi and Atchafalaya Rivers," *Estuaries* 16:541-546 (1993).

38. Flegal, A.R., G.J. Smith, G.A. Gill, S. Sanudo-Wilhelmy and L.C.D. Anderson, "Dissolved Element Cycles in the San Francisco Bay Estuary," *Mar. Chem.* 36:329-363 (1991).

39. Paulson, A.J. and R.A. Feely, "Dissolved Metals in Surface Waters of Puget Sound," *Mar. Pollut. Bull.* 16:285-291 (1985).

40. Hanson, A.K., Jr. and J.G. Quinn, "The Distribution of Dissolved and Organically Complexed Copperand Nickel in the Middle Atlantic Bight," *Can. J. Fish. Aquatic Sci.* 40:151-161 (1983).

41. Klinkhammer, G.P. and M.L. Bender, "Trace Metal Distributions in the Hudson River Estuary," *Est. Coast. Shelf Sci.* 12:629-643 (1981).

42. Nixon, S.W., "Metal Inputs to Narragansett Bay: A History and Assessment of Recent Conditions," Rhode Island Sea Grant Communications Office, (Order# P1388), 183 pgs. University of Rhode Island, Bay Campus (Narragansett, RI 1995).

DETERMINATION OF METAL SPECIATION IN SEAWATER USING CATHODIC STRIPPING VOLTAMMETRY

Constant M.G. van den Berg
Oceanography Laboratories
University of Liverpool
Liverpool, L69 3BX
United Kingdom

1. INTRODUCTION

The solubility of metals in natural waters is increased above that of the free (uncomplexed) metal ion by complexation with ligands (hydroxide, chloride, carbonate, sulfate and other anions) naturally occurring in the water [1]. In the seventies, it was proposed that metal concentrations in lake and oceanic basins were controlled by adsorption reactions which could be modeled by comparatively simple chemical models [2] in which metal scavenging is controlled by adsorption sites on particles of chiefly inorganic origin. With improved sampling and measuring techniques from the late seventies and eighties onward, it has become known that the metal concentrations in waters from unpolluted lakes and ocean basins are lower than expected and also display unexpected geochemistries. The distribution pattern of metals in the oceanic water column frequently is not typical of that expected for scavenged elements (concentration decreasing with depth): for instance the oceanic distribution of nickel is more like a nutrient (removal in the upper water column, regeneration in deep waters) and copper sometimes shows a gradual almost constant increase with depth [3]. The estimated oceanic residence time of 1000 years for copper [3] and longer for nickel, does not tally with an element that is scavenged quickly from solution. It is interesting therefore that recent measurements of the chemical speciation of copper have shown that it occurs fully complexed by organic matter in the seawater from oceanic origin [4,5], and also in estuaries [6] and lakes [7,8].

Organic complexes of metals behave differently from inorganic species. The adsorption on inorganic surfaces, such as on manganese oxides, is largely prevented by organic complexation, and the metal behaves more like the organic matter with which it is associated; it behaves much more conservatively. This explains, for instance, the largely conservative behavior of copper and zinc in estuarine waters outside the turbidity maximum [6].

Organic complexation is also of importance to toxicity and bioavailability of metals to microorganisms. It is known from culture experiments that the availability of various trace metals can be controlled by varying the concentration of a chelating agent (EDTA) [9], and that for instance the copper toxicity is diminished by organic ligands released by algae [10].

The chemical speciation therefore is fundamental to the biogeochemistry of trace elements in the aqueous system, and it is an important parameter to be determined to evaluate environmental impacts. Such measurements have so far not been part of the repertoire of routine analyses in laboratories involved in monitoring of the environment because of the perception that such measurements are complicated and time consuming, which indeed they were until recently. Electrochemical techniques are uniquely suited to determine metal speciation as the response is species related. Recent developments enable automation of the determination of metal speciation which should make such techniques more generally available. Here the use of voltammetry will be discussed to determine organic metal complexation in seawater.

2. THEORY: DISSOLVED METAL SPECIES

The various metal species in seawater, and their relationships, are shown in Table 1. The main species are M', ML and L'. The metal-organic speciation in a water sample would be known if M' and ML were determined directly. The metal speciation can also be calculated from the total metal and ligand concentrations (C_M and C_L) and the conditional stability constant (K'_{ML} or K''_{ML}): in this case the actual species have not been determined but are evaluated from ligand concentrations and stability constants which have to be determined for each sample separately; with this information a value for the free metal ion concentration (pM) can be calculated. The two methods are discussed below.

Table 1. Relationships between metal species in water.

Symbols and Metal Species	
M'	free inorganic metal (not complexed by organic ligands)
C_M and C_L	total metal and ligand concentrations
ML	organic complexes of M
L'	free ligand (not complexed by M)
Relationships	
$K'_{ML}=[ML]/[M^{n+}][L']$	$K''_{ML}=[ML]/[M^{n+}][L']$
Relationship between M^{n+} and M': $[M']=\alpha_M[M^{n+}]$	Relationship between M^{n+} and C_M: $[M^{n+}]=C_M/(\alpha_M+\alpha_{ML})$
Relationship between K'_{ML} and K''_{ML} $K'_{ML}=\alpha_M K''_{ML}$	$\alpha_M=1+((K_{MXin}[X_i]^n)$ where X_i=inorganic anion (Cl^-, HCO_3^-, OH^-)
$\alpha_{ML}=K'_{ML}[L']$	$\alpha''_{ML}=K''_{ML}[L']$

3. METHODS AND EXAMPLES

3.1 Actual Species Detection (M′ or ML)

The actual species occurring in seawater are difficult to determine directly (without prior separation and preconcentration) due to the very low concentration of the free metal ions. Determinations of the free metal ion concentration by potentiometry using an ion-selective electrode have so far been doomed to failure due to the low total metal concentrations combined with slow kinetics of organic interactions. Attempts at the cupric ion concentration in pH 2 and pH 8 seawater [11] have been inconclusive.

3.2 Anodic Stripping Voltammetry

Anodic stripping voltammetry (ASV) is potentially more successful as it is more sensitive, but is dogged by problems related similarly to the often vanishingly low levels of reactive metal species and differently due to changes in the speciation during the measurement that cannot be accounted for. In ASV the metal is first plated onto a mercury electrode and its concentration is subsequently evaluated from the oxidation current when the metal is stripped back into solution by scanning the potential on the electrode towards more positive values. The "labile" metal concentration is that which is plated onto the electrode.

The reduction potential of the complexed metal (ML) is shifted in a negative direction with respect to the uncomplexed (by L) metal. Careful selection of a deposition potential just negative of the reduction potential of the inorganic species should enable differentiation between organic and inorganic species. However, the peak shift for a two-electron reduction is only 30 mV/decade of α_{ML}, and a comparatively large overpotential of ca 200 mV is required to achieve optimal deposition of the metal onto the electrode. Only very stable complexes with a value for $\alpha_{ML}/\alpha_M>10^7$ would therefore be detected while much weaker complexes with $\alpha_{ML}/\alpha_M>1$ would already be significant in terms of complexation ability.

An exception is formed by organic complexes which dissociate slowly--inert species. Complexes that do not dissociate during the time spent in the diffusion layer at the electrode during the plating step are kinetically inert and are not detected. Experiments using a hanging mercury drop electrode have shown that the complex dissociation is significant for organic complexes in seawater causing significant underestimation of the degree of organic complexation [12]. The residence time in the diffusion layer is much reduced when a mercury film electrode is used in combination with either a rotating disk or a jet. In this case the dissociation of organic metal species becomes a function of the thickness of the double layer, the kinetics of complex dissociation, and the deposition potential. There is no evidence that the dissociation can be made to be insignificant, so some dissociation has to be expected and the labile metal fraction can therefore not be equated with the inorganic metal concentration (M′): The labile metal concentration includes inorganic and weaker organic complexes while the non-labile fraction comprises very stable or kinetically inert metal species.

3.3 Cathodic Stripping Voltammetry with Ligand Competition

Cathodic stripping voltammetry (CSV) is preceded by adsorption of a complex of the metal with a specific added ligand on the surface of the electrode (normally an HMDE) which is shown diagramatically in

Figure 1. The adsorbed layer is reduced during a negative going potential scan. The deposition normally is carried out at oxidative potentials so there is no change in the oxidation state during the deposition step.

Principles of adsorptive cathodic stripping voltammetry

1) adsorption of complex on electrode surface
2) reduction of element or ligand in the adsorbed complex

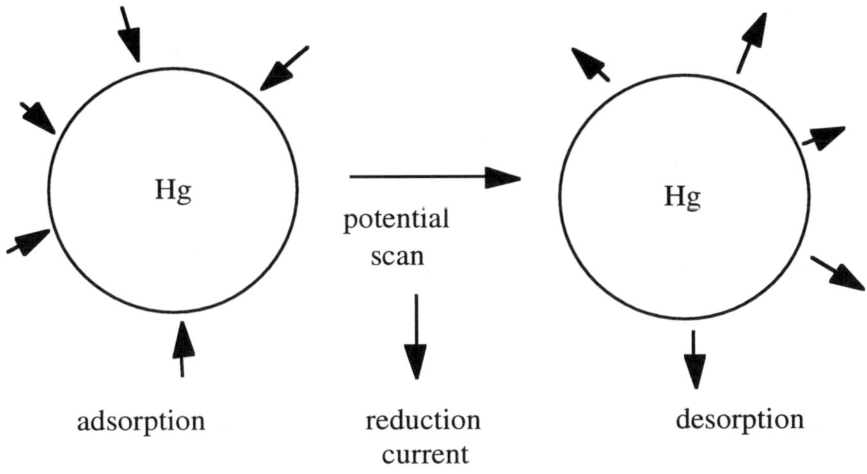

Requirements:
formation of adsorptive complex
presence of reducible groups

***Figure 1.** Diagram Illustrating the Principles of CSV Preceded by Adsorptive Deposition of Metal Complexes on the Surface of a Mercury Drop Electrode.*

The speciation in the seawater is perturbed by the addition of the active ligand (AL) (see Table 2); the magnitude of the alteration depends on the stability of the MAL complexes, and can be regulated by varying the concentration of AL or by selecting ligands of varying stabilities. Thus it is possible to investigate selectively strong or weak natural complexes by varying the "detection window." The labile metal concentration in CSV therefore equals all metal bound by the added ligand, which comprises all metal complexes weaker than those with AL. The distribution of metal species depends on the relative stabilities of the complexes (Table 2), so the labile metal concentration can be made to approach the total metal concentration by using a chelator forming very stable complexes and thus outcompeting the natural species.

Measurements of the labile metal concentration are carried out in equilibrium conditions if the added chelator is allowed to equilibrate with the metal species (this can take several hours). The concentration of MAL is normally a significant fraction of the total dissolved metal concentration (typically 50% is complexed by MAL) and only a very small amount is adsorbed on the electrode, so no changes in the speciation occur during the actual measurement of MAL. The measurement is therefore independent of kinetic effects of metal species up to time-scales of hours.

Table 2. Theory of metal speciation using ligand competition with CSV: labile metal concentrations.

Addition of AL causes a redistribution of metal species:

$$MAL \xleftarrow{\alpha_{MAL}} M^{n+} \xrightarrow{\alpha_{ML}} ML$$

where $\alpha_{MAL} = K'_{MAL}[AL']$

Operational definition of the labile metal concentration:

[labile metal] = [MAL] + [M′] and $\alpha_{ML} = \alpha_{MAL}$ (valid for the non-labile fraction)

Relationship between the labile and total dissolved metal concentrations:

[labile metal] = $C_M\ \alpha_{MAL}/(\alpha_{MAL} + \alpha_{ML})$

Calculation of α_{ML} from the dissolved and labile metal concentrations (valid if $C_L > C_M$):

$\alpha_{ML} = (C_M/[\text{labile metal}] - 1)\ \alpha_{MAL}$

Calculation of metal speciation in the original seawater prior to the addition of AL:

$[M'] = C_M / (\alpha_M + \alpha_{ML})$ and $[ML] = C_M - [M']$

The sensitivity of CSV is very high as the adsorbed metal species are collected in a single molecular layer which is reduced in its entirety during the scan. A metal is detectable if its complex with a suitable chelating agent is adsorbed on the surface of a mercury drop electrode, and if either the metal or the ligand moiety of the complex is reducible to a lower oxidation state. Some twenty metals are now detectable by CSV in seawater, while several other metals are detectable only when present at higher than natural concentration or after a suitable preconcentration step. Figure 2 shows which elements in the periodic system can be determined by CSV.

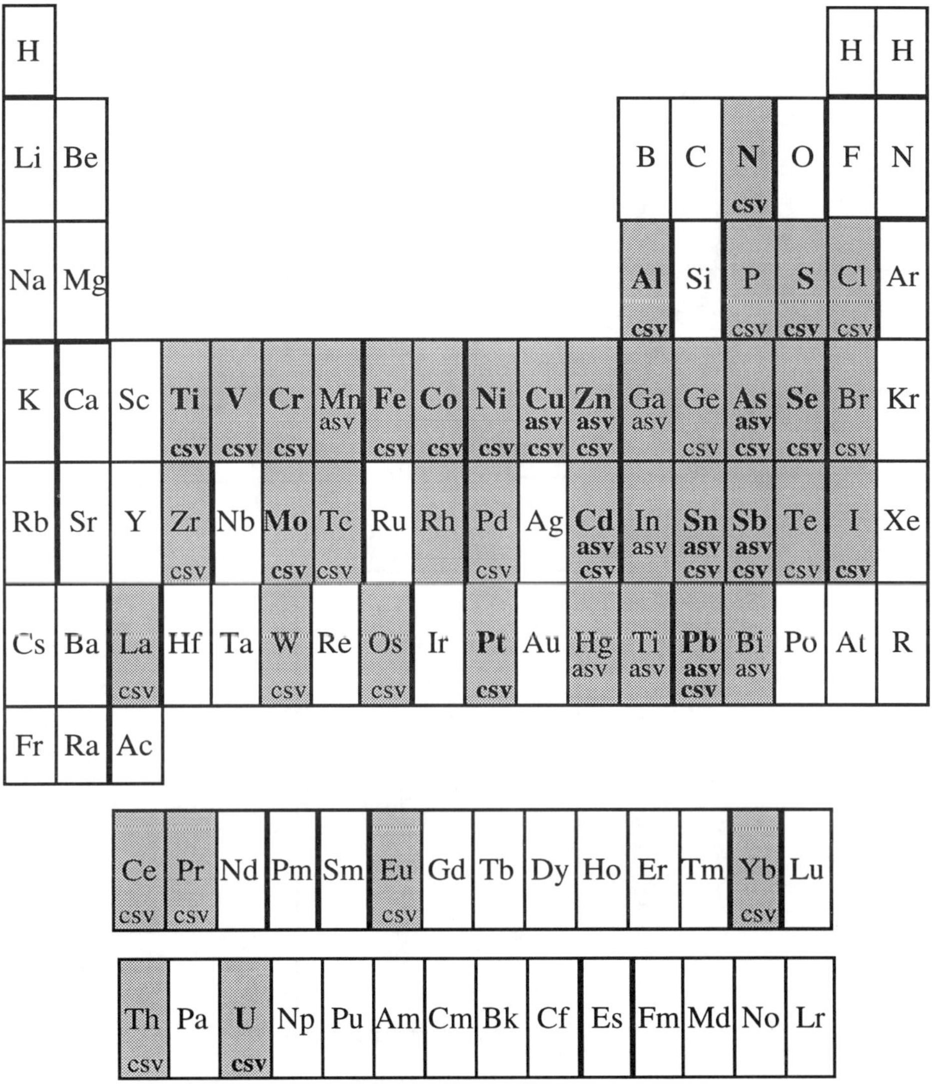

Figure 2. ***Diagram Showing Which Elements Can Be Determined by CSV in Natural Waters. Elements that can be determined directly in sea water are indicated in bold letters.***

Example CSV scans for one metal (3 nM copper in seawater) are shown in Figure 3. Several different ligands can be used to determine copper in seawater, best sensitivity being obtained using salicylaldoxime [13]. The availability of several ligands allows us to choose one optimal for the stability of the natural complexes occurring in the water being studied. Although best sensitivity is obtained using salicylaldoxime, the sensitivity of any of these ligands is sufficient to detect low copper levels in sea water as the sensitivity can be enhanced by extending the adsorption time. Typically the sensitivity increases linearly with the adsorption time between 1 and 10 min.

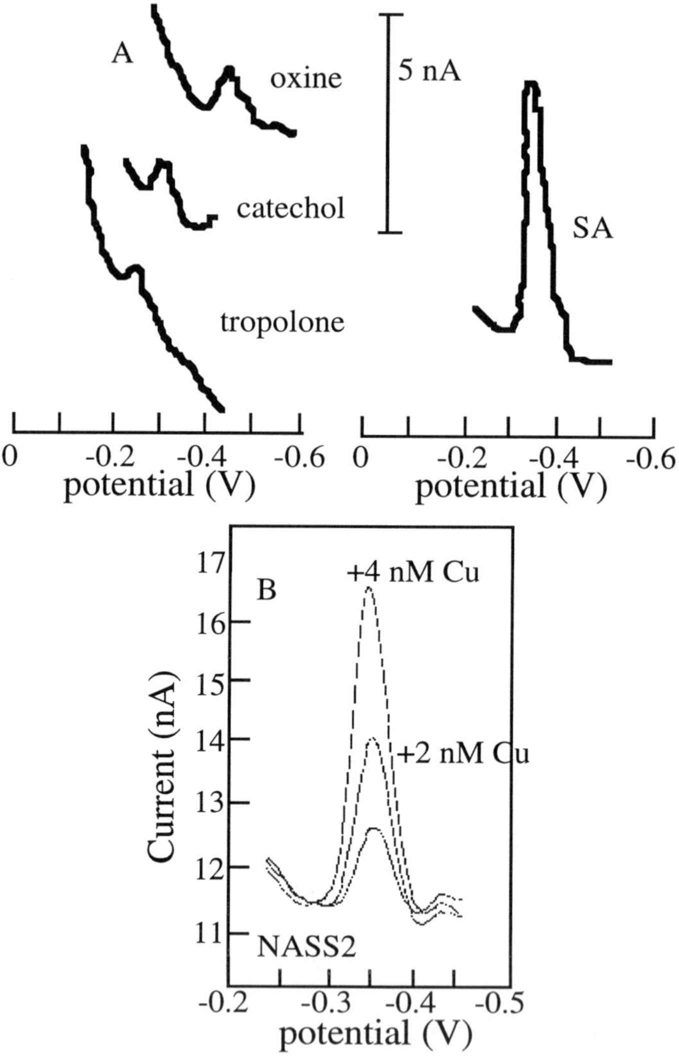

***Figure 3.** Comparative Scans for 3.1 nM Copper in Seawater Using CSV with Complex Adsorption with Oxine, Catechol, Tropolone and Salicylaldoxime (SA). Scans for copper in certified sea water (NASS2) are shown on the right with and without additions of 2 and 4 nM copper.*

3.4 Examples of Metal Speciation Using Labile and Total Dissolved Metal Determinations: Ti, Al, Cr, Fe

The most simple method of speciation is to differentiate between labile and total dissolved metal fractions by using UV-digestion to remove dissolved organic matter. Knowledge of the stability of the complexes with the added ligand allows an estimate of the degree of complexation of the metal but this is valid only if the natural ligand concentration is greater than that of the metal (Table 2). Examples will be given here for titanium, aluminum, nickel, chromium and iron.

Titanium is determined by CSV in the presence of mandelate and chlorate at pH 3 in sea water [14]. The sensitivity of this method is very high as advantage is taken of a catalytic effect of titanium on the reduction of chlorate; the limit of detection for titanium is ca. 7 pM. The concentration of titanium in sea water is quite low at around 0.1 nM where it occurs as titanium(IV). The inorganic speciation of titanium is dominated by $TiO\ (OH)_2^0$ with a value for α_{Ti} of 10^{11} [15]. Labile titanium concentrations in the water column of the Mediterranean are shown in Figure 4. Interestingly, less than half of the dissolved titanium is labile to the CSV measurements, the remainder being released by UV-digestion of the samples. Heat treatment of the samples to 70°C (similar to the temperature achieved during UV-digestion) did not cause the titanium to become released, suggesting that organic matter rather than inorganic colloidal matter was responsible for the non-lability of the titanium [15], indicating that some very stable organic complexes of titanium must exist in sea water.

The determination of aluminum by CSV is an example where use is made of reduction of the ligand in the adsorbed complex as the aluminum itself is reducible only at very negative potentials. The aluminum complex with dihydroxyanthraquinone is adsorbed on the electrode, and the reduction current of the ligand in the complex is directly related to the labile aluminum concentration in solution. It is perhaps not expected that aluminum reacts with dissolved organic matter in sea water where it occurs chiefly as $Al\ (OH)_3^0$ and $Al(OH)_4^-$ with a value for α_{Al} of $10^{8.9}$. Nevertheless, comparison of labile and total dissolved aluminum concentrations in the Mediterranean (Figure 4) shows that a significant fraction of the aluminum (close to 80%) is non-labile. It is not clear whether the non-labile fraction is inorganically colloidal or organically complexed (or colloidal) because, while this fraction is released upon UV-digestion at reduced pH (pH 2), it is also largely released upon prolonged acidified storage.

These two examples show that CSV can be used successfully to determine the presence of labile and non-labile species of individual metals, sometimes occurring at very low concentrations. The interpretation of the data is not always straightforward as it is normally necessary to carry out additional experiments to verify whether the non-labile species is of organic or inorganic nature. One way of doing this is to carry out titrations of the sea water with the metal to see whether more of the non-labile species can be formed. Comparison with UV-digested aliquots can be used to demonstrate the organic nature of the ligands, and the concentration and stability of the complexes can be determined (see below).

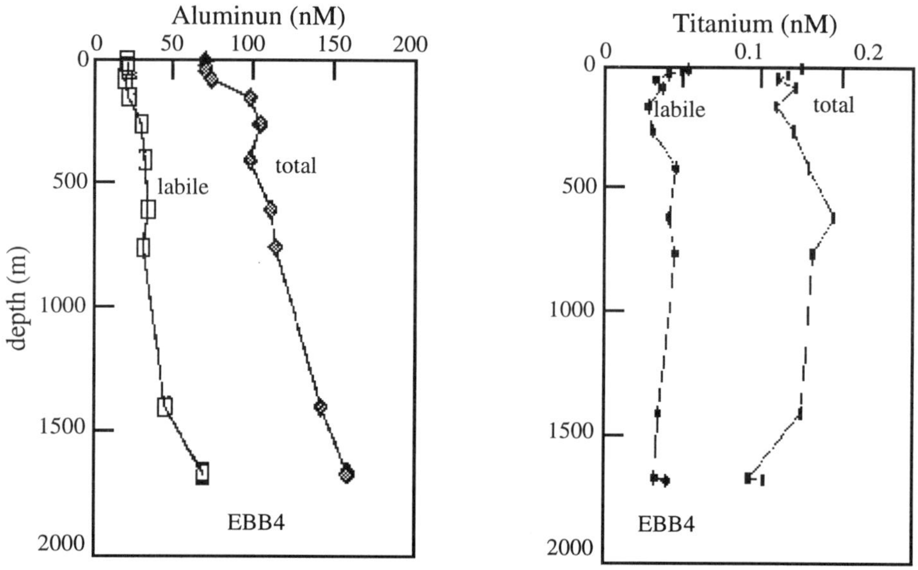

Figure 4. Speciation of Titanium and Aluminium in the North Western Mediterranean (data from van den Berg et al. [15]). Measurements by CSVat pH 3 (titanium) and pH 7 (aluminum).

3.5 Determination of Ligand Concentrations and Conditional Stability Constants Using CSV

More or less of the complexed metal can be exchanged by an added competing chelating ligand by varying the concentration of the added ligand or by using ligands forming complexes of differing stabilities. Assuming for the moment that the labile fraction of a metal equals its non-labile fraction in a sea water sample (i.e., half of the metal is labile), it can be understood intuitively that the metal-complex stability of the unknown complex equals that with the added ligand; $\alpha_{ML}=\alpha_{MAL}$. This comparison of complex stabilities forms the basis of ligand competition. A value for α_{MAL} is calculated from the product of K'_{MAL} and [AL′]; here K'_{MAL} is a conditional stability constant valid for pH 8 sea water (this involves correction for side-reactions with the major cations in sea water and pH, and for the ionic strength of sea water).

3.5.1 Calibration of K'_{MAL}

This constant can be calibrated (if no value is available) by comparison with another known competing ligand, usually EDTA for which the complex stabilities with many metals as well as the major ions in sea water are known at several ionic strengths and temperatures (e.g., [16]) and which does not form adsorptive complexes in CSV. The effect of adding EDTA to sea water is to lower the CSV response corresponding with the ligand competition exerted between the two ligands. The calibration of the complex stability of salicylaldoxime with copper in sea water is shown in Figure 5. The ratio of i_p/i_0 (the current in the presence divided by the current in the absence of EDTA) drops with increasing EDTA concentration. The parallel curves were obtained at different salinities. Values for the conditional stability constants are shown plotted against the (log) salinity on the right-hand side of Figure 5. These conditional stability constants can be used to calculate complex stabilities of unknown complexes of the type ML in sea water when the known ligand AL is used to determine unknown ligands in natural waters by ligand competition.

3.5.2 Complexing ligand titrations

Addition of a competing ligand AL to a sample causes some of the metal initially complexed by L to become labile by complexation with AL until [labile metal] = $C_M\ \alpha_{MAL}/(\alpha_{MAL} + \alpha_{ML})$ (Table 2). Thus part of initially complexed L (by M) is liberated. Titration of a ligand L with metal ions while monitoring the labile metal concentration with CSV causes the response to increase gradually and non-linearly until the ligand is saturated, whereafter a linear response is obtained. Thus, first the ligand is titrated which was freed upon addition of AL, and subsequently any excess ligand. The total ligand concentration is determined. The curved response is confirmation of the presence of a finite concentration of complexation sites.

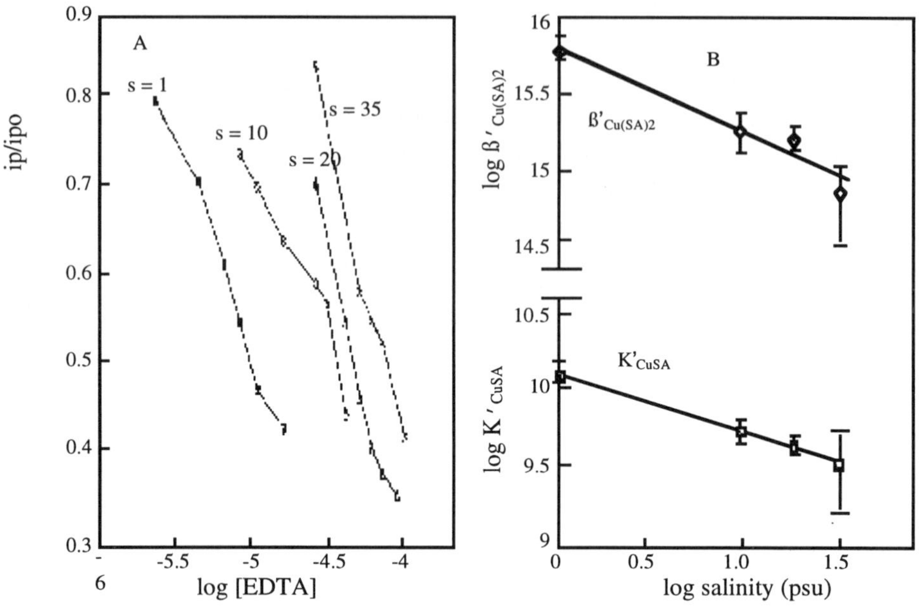

Figure 5. Calibration of the Conditional Stability Constants for Complexation of Copper by Salicylaldoxime in Sea Water by Ligand Competition Against EDTA (data from Campos and van den Berg [13]). Lowering of the peak height by EDTA additions (A), and values for K'_{CuSA} and α'_{CuSA2} (B).

An example of this effect is shown for a titration with iron of a sea water sample originating from a depth of 50 m in the N.W. Mediterranean, while monitoring the labile iron concentration by CSV in the presence of nitrosonaphthol (NN) (Figure 6a): the peak height for iron increases non-linearly until the natural ligand(s) have been saturated, whereafter the increase in the peak height is linear with the iron concentration. Linearization of the data (Figure 6b and Table 3) is used to calculate the ligand concentration C_L=9.5 nM and the conditional stability constant (log K'_{FeL}=21.5). The dissolved iron concentration in this sample was ~3 nM, well below the ligand concentration. Furthermore the stability of the complex is very great with a value for log α_{FeL}=13.5, much greater than that for inorganic complexation of iron, log α_{Fe}=11.3 at pH 8. All iron therefore occurred fully complexed by organic matter in this sample in spite of the very great stability of the inorganic (hydroxide) species.

Table 3. Theory: Complexing ligand titrations using ligand competition with CSV.
A. Calibration of α_{MAL} by ligand competition against a known complex like EDTA: $\alpha_{MAL} = [(\alpha_M + \alpha_{MEDTA})X - \alpha_M] / (1 - X)$ where $X = i_0/i_p$ (see Figure 6a).
B. Data treatment of labile metal concentrations resulting from a ligand titration with metal ions: $[\text{labile metal}]/[ML] = [\text{labile metal}]/C_L + (\alpha_M + \alpha_{MAL})/(C_L K'_{ML})$
Values for C_L and K_{ML} can be calculated from respectively the slope, $(C_L)^{-1}$, and the Y-axis intercept, $(\alpha_M + \alpha_{MAL})/(C_L K'_{ML})$, of a linear least squares regression of [labile metal] as a function of [labile metal]/[ML] [17, 18], or from a non-linear data treatment (e.g., [19]).

Investigations into the chemical speciation of iron in the oceanic water column using CSV with ligand competition have now shown that iron is fully and very stably complexed by organic matter throughout the water column (Figure 6c) [20-22]. This is a very important finding in view of the known limitation of oceanic primary productivity by lack of iron, suggesting that the microorganisms may be altering their environment to maintain iron in solution and perhaps in a form that is especially suitable for uptake. The current speculation is that siderophores are responsible for the stable organic complexation of iron. The distribution of organic iron complexing ligands in the water column of the Mediterranean is shown in Figure 6c along that of dissolved iron. The ligand concentration increases at depths at and below the maximum in the primary productivity (not shown here) suggesting that marine microorganisms (phytoplankton or bacteria) are responsible for the ligands. The ligand concentration is always higher than the iron concentration; the high stability of the complexes causes >99% of the iron to occur organically complexed by these ligands.

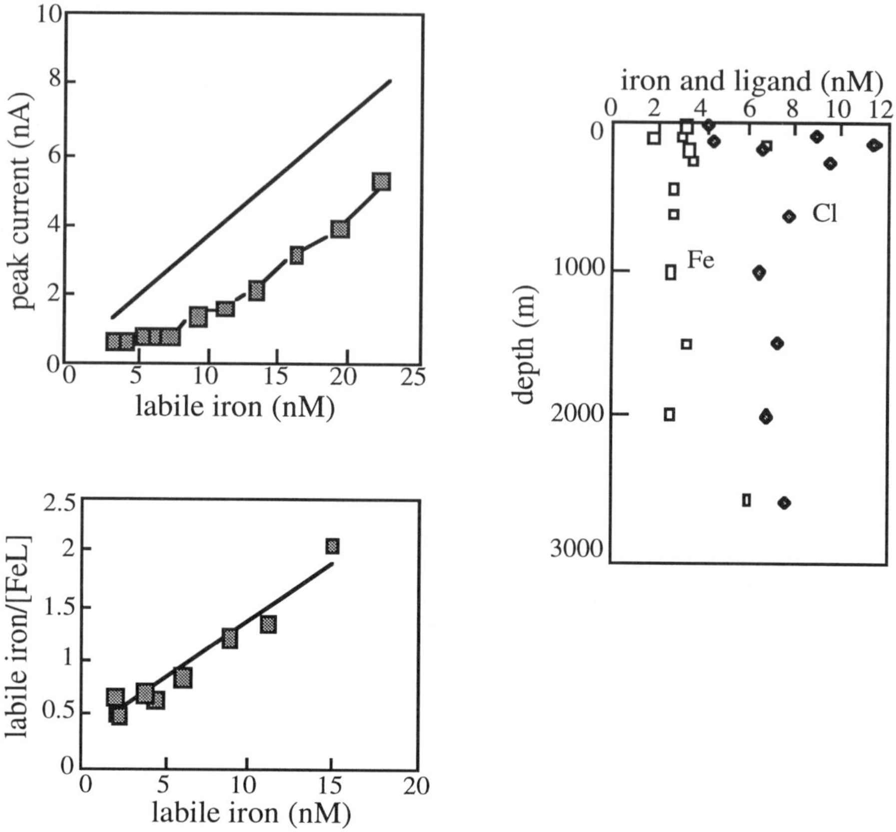

Figure 6. ***CSV Labile Iron During a Titration with Iron of a Sample from 50m Depth in the NW Mediterranean. a: labile iron concentrations showing a non-linear increase until the natural ligand has been titrated; b: linearization of the data to calculate the ligand concentration (C_L=9.5 nM) and the conditional stability constant (log K'_{FeL}=21.5) (data from van den Berg [20]). c: the distribution of iron complexing ligands in the NW Mediterranean (data from van den Berg [20]).***

Separate investigations have shown that copper, zinc and nickel occur to various extent complexed in sea water. It has been reported variously that copper is organically complexed chiefly in the upper water column [4] and throughout the water column [5]; zinc is probably organically complexed throughout the water column [23]; nickel may be fully organically complexed but the ligand used for the determination outcompetes the natural complexes so only low concentrations of complexing ligands are detected which bind a small fraction (10-20%) very stably [24]. It appears that several complexing ligands are present in the oceanic water column which bind metals with varying stabilities; the detection of each ligand depends on the metal and on the "detection window" of the analytical technique [25,26]. The detection window comprises the metal complexes with values for α_{ML} within a decade of α_{MAL}; complex stabilities outside this range are not detected because the metal complexed either does not dissociate or no labile metal is detected.

3.6 Determination of Redox Speciation: Cr and Fe

It can be difficult to determine redox speciation by voltammetry if the redox reactions are reversible at the electrode surface. Some redox species are not electroactive if the electrochemical reduction is kinetically inhibited (for instance, arsenic(V) and selenium(IV) are not readily reduced at the electrode surface) so only their counterparts are detected; in several cases redox species can be stabilized using selective chelation Fe(II) stabilization with dipyridyl [27]) or separated prior to the measurement (e.g., chromium(III) [28]). Thus arsenic(III), selenium(III) and iodide are determined selectively by voltammetric methods while total arsenic, selenium and iodine are obtained only after suitable sample treatment. Chromium and iron are discussed here in more detail.

Chromium (VI) is detected by CSV in the presence of DTPA [28,29]. The electrochemistry is quite interesting. Chromium(VI) is reduced to chromium(III) at the deposition potential (-0.8 V) at the electrode surface; the freshly produced chromium(III) forms a complex with DTPA, adsorbs on the electrode, and is reduced to chromium(II) in the subsequent potential scan. Interestingly, the chromium(II) catalyzes the reduction of nitrate causing a large reduction current. Thus, the sensitivity for chromium is greatly enhanced by employing this catalytic effect by measuring in the presence of nitrate at pH 5 to 6; the limit of detection of chromium in sea water is ~0.1 nM after a deposition time of 2 min [28]. Usefully, the CSV sensitivity for chromium in freshwater is 10 times higher than in sea water. Scans for chromium(VI) in sea water are shown in Figure 7a. Dissolved chromium(III) that is initially present in the sample also contributes to the reduction current, but its contribution rapidly decays due to the gradual formation (completed within ca. 30 min) of an inert species which then no longer participates in the reactions. A convenient method to remove chromium(III) prior to the determination of chromium(VI) is to add a sorbent of silica particles to the solution to adsorb all chromium(III) prior to the addition of the ligand and the CSV determination of chromium(VI). Total dissolved chromium is determined after UV-digestion of the sample which causes all chromium(III) to be converted to chromium(VI) [27].

The CSV method has been used to determine the redox speciation of chromium in the water column of the Mediterranean. The vertical distribution of chromium(III) and (VI) is shown in Figure 7b. Typically most chromium occurs in the VI oxidation state as expected for thermodynamic reasons. However, unexpectedly a large fraction (up to ~1 nM out of a total of ~3.5 nM) of the chromium occurs as the (more toxic) chromium(III) in the upper water column and in deep waters [24]. The reasons for this phenomenon are being debated and center on photochemical and biological effects.

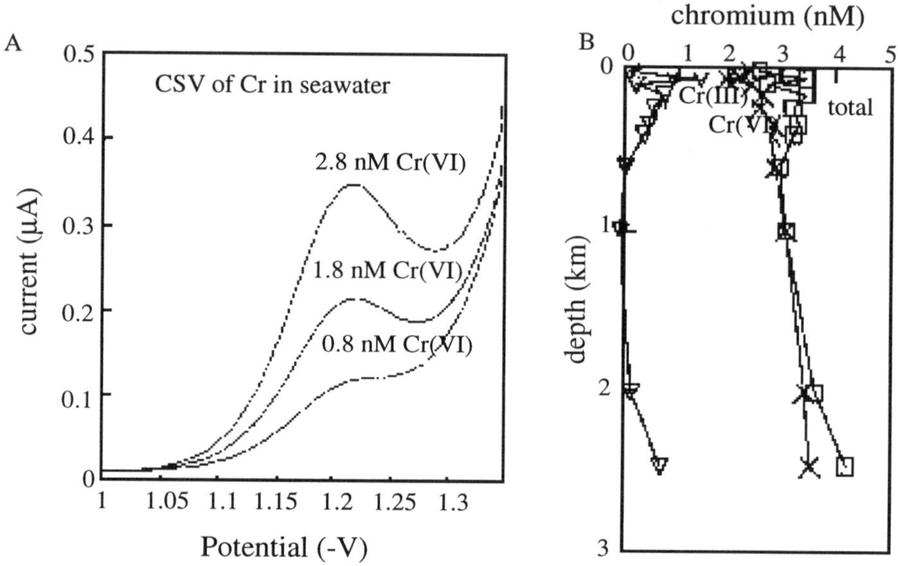

Figure 7. ***CSV of the Redox Speciation of Chromium in Sea Water. a: CSV scans for chromium(VI). The following reagents were added to the sea water: 2.5mM diethylenetriaminepentaacetic acid (DTPA), 0.01 M acetate buffer of pH 5.2 and 0.5 M nitrate. The deposition time on the mercury drop electrode was 2 min. b: vertical profiles for chromium(III) and (VI) in the water column of the Mediterranean (data from Achterberg and van den Berg [24]).***

Iron occurs as either iron(II) or (III) in natural waters, e.g., [30]. Its reduction is mediated by photochemical reactions [31] and by activities of microorganisms. It is important to know the redox speciation in natural waters as it affects the availability to microorganisms. Low concentrations of iron can be determined by CSV in the presence of nitrosonaphthol as complexing ligand. The sensitivity is greatly enhanced in the presence of hydrogen peroxide as its reduction is catalyzed by iron(II). This is the basis for a catalytic method where sub-nanomolar levels of iron are determined by CSV after a deposition period of one minute [32]. A problem for the determination of the redox speciation is that both iron(III) and (II) respond equally well. Fortunately, iron(II) can be complexed selectively by adding dipyridyl and is then effectively masked from the CSV determination. Thus iron(III) is determined selectively by CSV and the concentration of iron(II) is obtained by difference between the total dissolved iron and iron(III) concentrations [27]. Scans for iron are shown in Figure 8a with additions of iron(II) in the presence (added iron(II) is masked) and absence of dipyridyl. Measurements of the distribution of iron(II) in the water column of the North Sea using this technique have shown that it constitutes an important fraction in the surface waters (probably as a result of photochemical effects) but also in deeper waters possibly by releases from porewaters (Figure 8b) [27]. It is interesting that the iron(II) persists for significant periods in the water in the presence of dissolved oxygen suggesting that it is probably produced continuously.

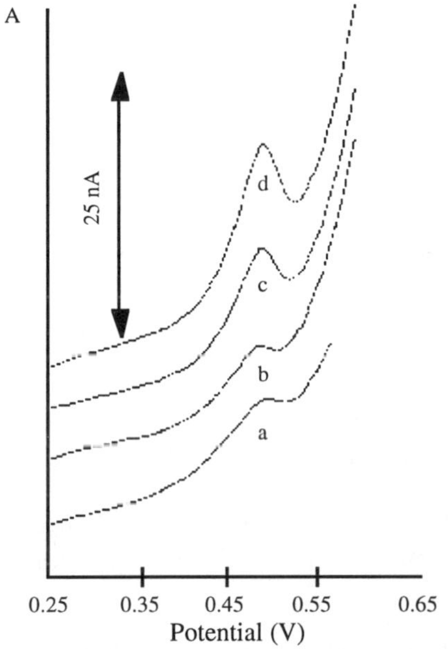

Figure 8. ***CSV of the Redox Speciation of Iron in Sea Water. a: CSV scans for iron in the presence of 25 μM nitrosonaphthol, 0.01 M PIPES pH buffer (pH 6.9), 1.8 mM hydrogen peroxide and 5 mg/L sodium dodecyl sulphate; iron(II) is masked by dipyridyl. Scans are shown for a) 1.8 nM Fe_R, b) after addition of 2.8 nM iron(II) and 2 M Dp (causing Fe(II) to be masked); c) as (a) but after addition of 2.8 nM iron(II) (no masking) and d) as (c) but after a further addition of 2.5 nM iron(III) (normal response for Fe(III)). b: distribution of the redox speciation of iron in the water column of the North Sea; data from Gledhill and van den Berg [27].***

4. CONCLUDING REMARKS

Various metal species, whether organic complexes or redox species, occurring in natural waters including sea water can be determined by the voltammetric techniques discussed here. The measurement is more complicated than a "simple" determination of the total dissolved metal concentration, but the added information is important as with it one can evaluate whether a given metal concentration is low or high with respect to toxicity to microorganisms; it is possible to obtain the important parameter pM by comparatively simple means. Recent developments in automation in sample preparation and analysis with voltammetric detection [33-35] mean that natural waters or effluents can be analyzed fully automatically using flow-analysis techniques. These procedures are still in their infancy and development is likely to continue to advance the capabilities of automated analysis including that of metal speciation in the near future.

REFERENCES

1. Stumm, W. and H. Bilinski, "Trace Metals in Natural Waters; Difficulties of Interpretation Arising from Our Ignorance on Their Speciation," *Advances in Water Pollut. Res.* pp. 39-49. Pergammon Press (London, 1973).

2. Schindler, P.W., "Removal of Trace Metals from the Oceans: A Zero Order Model," *Thalassia Jugoslavica* 11:101-111 (1975).

3. Bruland, K.W., "Trace Elements in Sea-water," in *Chemical Oceanography*, J.P. Riley, Ed., Vol. 8, pp. 157-220. Academic Press (London, UK 1983).

4. Coale, K.H. and K.W. Bruland, "Spatial and Temporal Variability in Copper Complexation in the North Pacific," *Deep-Sea Res.* 47:317-336 (1990).

5. Buckley, P.J.M. and C.M.G. van den Berg, "Copper Complexation Profiles in the Atlantic Ocean," *Mar. Chem.* 19:281-296 (1986).

6. van den Berg, C.M.G., A.G. Merks and E.K. Duursma, "Organic Complexation and Its Control of the Dissolved Concentrations of Copper and Zinc in the Scheldt Estuary," *Est. Coast. Shelf Sci.* 24:785-797 (1987).

7. van den Berg, C.M.G. and J.R. Kramer, "Determination of Complexing Capacities and Conditional Stability Constants for Copper in Natural Waters Using MnO_2," *Anal. Chim. Acta* 106:113-120 (1979).

8. Xue, H.B. and L. Sigg, "Free Cupric Ion Concentration and Cu (II) Speciation in a Eutrophic Lake," *Limnol. Oceanogr.* 38:1200-1213 (1993).

9. Brand, L.E., W.G. Sunda and R.R.L. Guillard, "Limitation of Marine Phytoplankton Reproductive Rates by Zinc, Manganese and Iron," *Limnol. Oceanogr.* 28:1182-1198 (1983).

10. van den Berg, C.M.G., P.T.S. Wong and Y.K. Chau, "Measurement of Complexing Materials Excreted from Algae and Their Ability to Ameliorate Copper Toxicity," J. *Fish. Res. Board Canada* 36:901-905 (1979).

11. Belli, S.L. and A. Zirino, "Behavior and Calibration of the Copper (II) Ion-selective Electrode in High Chloride Media and Marine Waters," *Anal. Chem.* 65:2583-2589 (1993).

12. van den Berg, C.M.G., "The Effect of the Deposition Potential on the Voltammetric Determination of Complexing Ligand Concentrations in Sea-water," *Analyst* 117:589-593 (1992).

13. Campos, M.L.A.M. and C.M.G. van den Berg, "Determination of Copper Complexation in Sea Water by Cathodic Stripping Voltammetry and Ligand Competition with Salicylaldoxime," *Anal. Chim. Acta* 284:481-496 (1994).

14. Yokoi, K. and C.M.G. van den Berg, "Determination of Titanium in Sea Water Using Catalytic Cathodic Stripping Voltammetry," *Anal. Chim. Acta* 245:167-176 (1991).

15. van den Berg, C.M.G., M. Boussemart, K. Yokoi, T. Prartono and M.L.A.M. Campos, "Speciation of Aluminum, Chromium and Titanium in the NW Mediterranean," *Mar. Chem.* 45:267-282 (1994).

16. Martell, A.E. and R.M. Smith, *Critical Stability Constants, Vol. 1: Amino Acids*, 469 pgs. Plenum Press (New York, NY 1974).

17. van den Berg, C.M.G., "Determination of Copper Complexation with Natural Organic Ligands in Seawater by Equilibration with MnO_2. I. Theory," *Mar. Chem.* 11:307-322 (1982).

18. Ruzic, I., "Theoretical Aspects of the Direct Titration of Natural Waters and Its Information Yield for Trace Metal Speciation," *Anal. Chim. Acta* 140:99-113 (1982).

19. van den Berg, C.M.G., "Determination of Complexing Capacities and Conditional Stability Constants of Complexes of Copper (II) with Natural Organic Ligands in Seawater by Cathodic Stripping Voltammetry of Copper-Catechol Complex Ions," *Mar. Chem.* 15:1-18 (1984).

20. van den Berg, C.M.G., "Evidence for Organic Complexation of Iron in Seawater," *Mar. Chem.* 50:139-157 (1995).

21. Rue, E. and K.W. Bruland, "Complexation of Iron (III) by Natural Organic Ligands in the Central North Pacific as Determined by a New Competitive Ligand Equilibration / Adsorptive Cathodic Stripping Voltammetric Method," *Mar. Chem.* 50:117-138 (1995).

22. Wu, J. and G.W. Luther, III, "Complexation of Iron (III) by Natural Organic Ligands in the Northwest Atlantic Ocean by a Competitive Ligand Equilibration Method and a Kinetic Approach," *Mar. Chem.* 50:59-177 (1995).

23. Bruland, K.W., "Complexation of Zinc by Natural Organic Ligands in the Central North Pacific," *Limnol. Oceanogr.* 34:269-285 (1989).

24. Achterberg, E.P. and C.M.G. van den Berg, "Chemical Speciation of Chromium and Nickel in the Western Mediterranean," *Deep-Sea Res.* (accepted, 1997).

25. van den Berg, C.M.G. and J.R. Donat, "Determination and Data Evaluation of Copper Complexation by Organic Ligands in Sea Water Using Cathodic Stripping Voltammetry at Varying Detection Windows," *Anal. Chim. Acta* 257:281-291 (1992).

26. van den Berg C.M.G., M. Nimmo, P. Daly and D.R. Turner, "Effects of the Detection Window on the Determination of Organic Copper Speciation in Estuarine Waters," *Anal. Chim. Acta* 232:149-159 (1990).

27. Gledhill, M. and C.M.G. van den Berg, "Measurement of the Redox Speciation of Iron in Seawater by Catalytic Cathodic Stripping Voltammetry," *Mar. Chem.* 50:51-61 (1995).

28. Boussemart, M., C.M.G. van den Berg and M. Ghaddaf, "The Determination of the Chromium Speciation in Sea Water Using Catalytic Cathodic Stripping Voltammetry," *Anal. Chim. Acta* 262:103-116 (1992).

29. Golimowski, J., P. Valenta and H.W. Nurnberg, "Trace Determination of Chromium in Various Water Types by Adsorption Differential Pulse Voltammetry," *Fres. Z. Applied Spectroscopy* 322:315-322 (1985).

30. O'Sullivan, D.W., A.K. Hanson, W.L. Miller and D.R. Kester, "Measurement of Fe (II) in Surface-Water of the Equatorial Pacific," *Limnol. Oceanogr.* 36:1727-1741 (1991).

31. Hudson, R.J.M., D.T. Covault and F.M.M. Morel, "Investigations of Iron Coordination and Redox Reactions in Seawater Using ^{59}Fe Radiometry and Ion-Pair Solvent Extraction of Amphiphilic Iron Complexes," *Mar. Chem.* 38:209-235 (1992).

32. Yokoi, K. and C.M.G. van den Berg, "The Determination of Iron in Seawater Using Catalytic Cathodic Stripping Voltammetry," *Electroanalysis* 4:65-69 (1992).

33. Achterberg, E.P. and C.M.G. van den Berg, "In-Line Ultraviolet-Digestion of Natural Water Samples for Trace Metal Determination Using an Automated Voltammetric System," *Anal. Chim. Acta* 291:213-232 (1994).

34. van den Berg, C.M.G. and E.P. Achterberg, "Automated In-Line Sampling and Analysis of Trace Elements in Surface Waters with Voltammetric Detection," *TrAC* 13:348-352 (1994).

35. van Elteren, J.T., C.M.G. van den Berg, H. Zhang, T.D. Martin and E.P. Achterberg, "Automated In-Line Extraction of U(VI) from Raffinate Streams with On-Line Detection by Cathodic Stripping Voltammetry," *Anal. Chem.* 67:3903-3910 (1995).

NEW DEPARTURES FOR ELEMENTAL SPECIATION STUDIES UTILIZING PLASMA MASS SPECTROMETRIC DETECTION

Joseph A. Caruso, Nohora P. Vela and Lisa K. Olson
Department of Chemistry
University of Cincinnati
Cincinnati, OH 45221

1. INTRODUCTION

Liquid chromatographic, gas chromatographic and supercritical fluid chromatographic methods (HPLC, GC, and SFC) have been used extensively for the separation and speciation of organometallic, metallic, and organic compounds [1-8]. Plasma source mass spectrometry (MS) provides a means of chromatographic detection at sub-ng to sub-pg levels of detection. In this way, much useful information about the specific nature of the substance of concern can be obtained. For example, it is well known that certain elemental forms are very toxic, such as arsenite, while others such as arsenobetaine are relatively innocuous. Thus, speciation information at the sub-ng levels or lower are critical to the risk assessment process. This paper summarizes recent chromatographic studies using plasma MS detection. HPLC has been utilized with a variety of chromatographic types including reversed phase chromatography and ion chromatography [1-6]. Most of the HPLC applications have utilized inductively coupled plasmas (ICP) as sources [1-6]. GC, using capillary columns in the normal and supercritical fluid modes (SFC) shows tremendous potential for femtomolar detection levels [1,2,4,7,8]. New departures in elemental speciation studies include low-power and low-flow plasmas that allow quantitative determination at ultra-trace levels plus provide fragment information for structural determination [9-13]. These may be coupled to microbore liquid chromatography or capillary electrophoresis (CE) for μL sample introduction [14-16].

2. PLASMA SOURCES FOR IDENTIFICATION AND QUANTITATIVE DETERMINATION

Chromatography coupled to mass spectrometry is a powerful tool for the separation, identification and quantitation of numerous chemical compounds. However, many compounds that are thermally labile or nonvolatile are not amenable to analysis by GC/MS. Thus, there is a clear need to do LC introduction with MS detection and several options have been explored to facilitate their interfacing. Most recent approaches have been oriented to obtain charged molecular species at atmospheric pressure [10,11]. Atmospheric pressure ionization (API) is advantageous for liquid sample introduction to the mass spectrometer since it reduces the introduction of large amounts of solvent or vapor into the vacuum system, facilitating the interface of conventional LC flow rates with MS [10,11]. The potential of using the microwave induced plasma as an ionization source to obtain structural information has been previously demonstrated for gaseous [9] and liquid samples [10, 11]. Results published by Olson *et al.,* [9] indicate that by varying the MIP power, it was possible to change the fragmentation pattern for several low-molecular weight hydrocarbons. Shen *et al.,* [10] describe the use of a low-power atmospheric-pressure microwave plasma as an ionization source for gaseous and liquid samples. They used this ionization source to obtain structural information for volatile, low molecular weight organic compounds. Particularly important is the fact that with the microwave induced plasma, MIP, operation may be shifted from a "bare elemental ion" configuration giving quantitative elemental detection, to one that gives many fragments. This would be comparable to electron impact sources having a high potential for structural analysis.

Our initial LC-MIP-MS study [11] demonstrates the effects of microwave power, sample gas flow rate and injector probe configuration in the production of fragment ions using perfluorotributylamine (PFTBA or FC-43) that is used as a calibration compound for organic MS. Liquid sample introduction to the MIP is obtained by using a direct injection nebulizer (DIN). This nebulizer provides a good aerosol and operates at typical microbore HPLC flow rates.

A VG Plasma Quad, conventionally used for elemental ICP-MS analysis, was modified by replacing the ICP source with the MIP cavity. The microwave system consisted of a 120 W microwave generator and a modified Beenaker microwave TM 010 cavity (92.5 mm i.d. and 2 cm. depth). An injector probe is inserted in the conventional MIP quartz tube that is used as a torch for MIP-MS experiments [10,11]. Three different injector probes were used in these experiments and they are described as follows: Probe A: stainless steel tube, 1/16" o.d.; Probe B: stainless steel tube, 1/16" o.d. with a ceramic tube inside (1/32" o.d.) and Probe C: stainless steel tube, 1/8" o.d. with a ceramic tube inside (1/16" o.d.). The combined effect of the MIP power and different injector probe configurations was evaluated for the production of several fragment ions. Results indicate that the mass spectrum changes considerably, favoring lower mass fragments as the power is increased. Figure 1 shows the dramatic differences in the spectra.

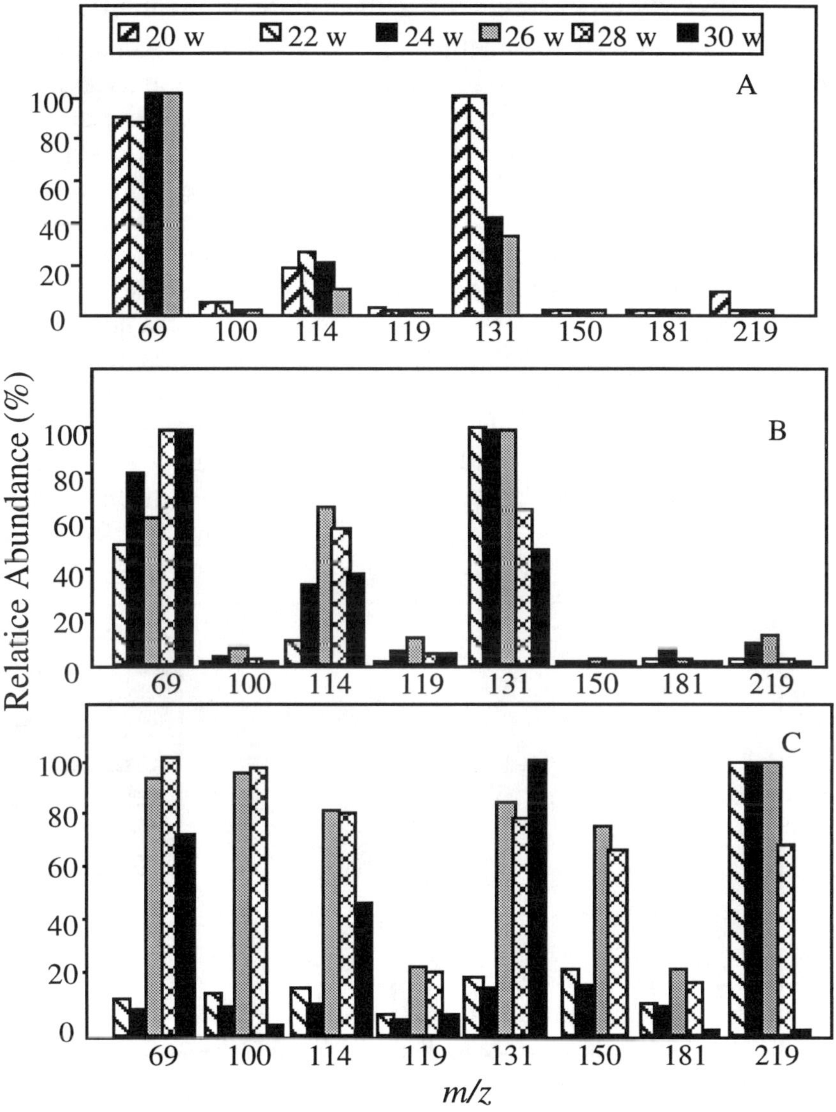

Figure 1. Effect of MIP power on the fragmentation of PFTBA. Gaseous sample introduction through injector probes A, B, and C. See text for injector probe description.

The use of a ceramic tube inside the stainless steel tube reduced the interaction between the plasma and the sample with a corresponding reduction in the extent of fragmentation at a given power. A larger diameter probe (injector probe C) indicated an increase in the abundance of all fragments. For powers between 22 and 26 W there is a shift to greater abundance of higher mass fragments. Ions representing molecular fragments larger than 250 a.m.u. cannot be monitored in these experiments due to the mass spectrometer's limited mass range. However, they would be seen with a quadrupole of greater mass range.

The spectrum for tetramethyl tin (M.W. 180) obtained at a MIP power of 24 W and presented in Figure 2 shows fragments at *m/z* 165, 150, 135 and 120. These correspond to successive methyl group loss. Near these fragments are the corresponding methylated fragments for the tin isotopes at *m/z* = 118 and 116. These fragments are at *m/z* =163, 161, 148, 146, 133 and 131. The results obtained in this study are in good agreement with the data published in the literature for the ionization of tetramethyltin. The literature reports the base peak at *m/z* = 165, corresponding to the trimethyltin ion. Other major fragments reported are $(CH_3)Sn^+$, $(CH_3)_2Sn^+$, SnH^+, $SnCH_2^+$. It is also important to note that the signal observed at *m/z* = 183 might correspond to the hydrated trimethyltin ion $[(CH_3)_3Sn{\cdot}H_2O]^+$, with the corresponding tin isotopes at *m/z* 181 and 179. Another hydrated ion, $[(CH_3)_3SnCH_2{\cdot}H_2O]^+$ is also detected at *m/z* = 197 with the analogous tin isotopes at *m/z* 191 and 193. Figure 2 also shows elemental tin and tin fragments at similar relative intensities.

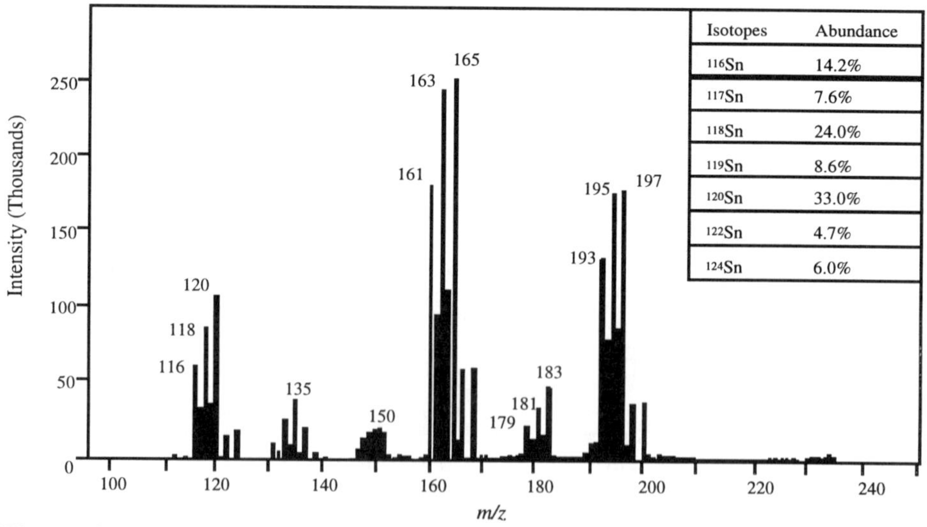

Figure 2. ***API Spectrum for Tetramethyltin, $(CH_3)_4Sn$. MIP incident power: 24 W; nitrogen carrier gas: 40 mL/min and helium gas: 5 L/min.***

The potential for using an atmospheric pressure low-power MIP as ionization source for liquid chromatography mass spectrometry was further demonstrated by using a 0.025 M solution of tetrabutyl ammonium hydroxide (M.W. = 259). The solution was introduced using a LC pump. Figure 3 shows the mass spectrum obtained for tetrabutylammonium hydroxide. The base peak is at *m/z* = 142 that corresponds to $(C_4H_9)_2C\text{-}NH_2^+$. The other dominant peaks in the mass spectrum of tetrabutyl ammonium hydroxide at *m/z* 184, 84, 86, 128 and 100 could be attributed to fragments such as $(C_4H_9)_2\text{-}N^+{=}CH\text{-}C_3H_7$, $(C_4H_9)\text{-}N^+{\equiv}CH$, $(C_4H_9)\text{-}NH^+{=}CH_2$, $(C_4H_9)\text{-}NH^+{=}CH\text{-}C_3H_7$, and $(C_4H_9)\text{-}NH^+{=}CH\text{-}CH_3$, respectively.

The stability of the plasma with aqueous solutions and the capability to obtain structural information demonstrates the potential for using this source for LC/MS studies. Also, the ability to switch modes from "bare" ion to molecular ion with fragments shows high potential for ultra-trace elemental and structural analysis.

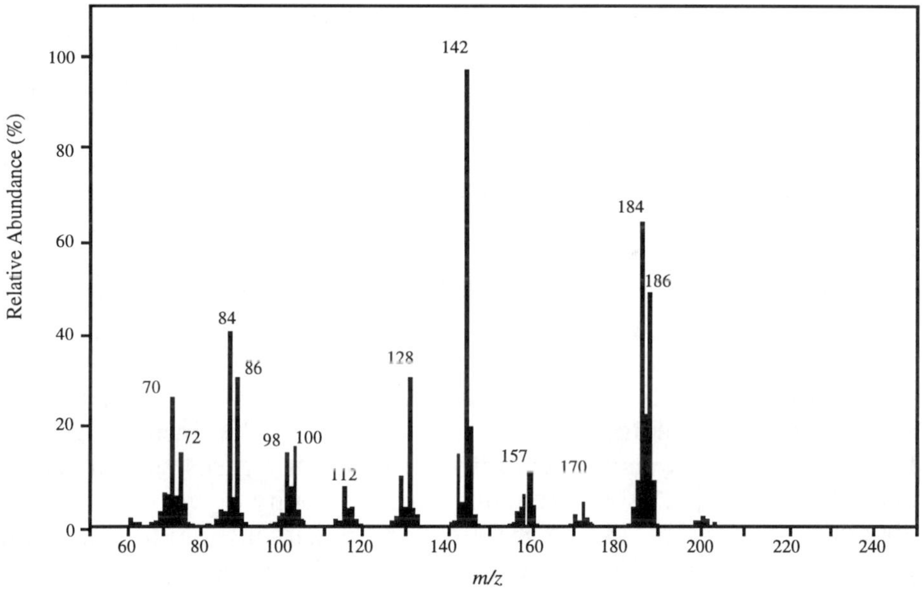

Figure 3. ***API Spectrum for Tetrabutyl Ammonium Hydroxide, MW=259. Liquid sample introduction using a LC pump with a liquid flow rate of 20 to 30 µL/min. MIP incident power: 24 w; nitrogen carrier gas: 40 mL/min and helium gas: 5 L/min.***

3. GC-GD FOR ELEMENTAL SPECIATION AND STRUCTURAL DETERMINATIONS AT SUB-NG TO SUB-PG LEVELS

Although glow discharge (GD) techniques currently are most applied to solid samples, it is of interest to investigate the usefulness of a particular plasma source for other sample types [12,13]. The glow discharge should be particularly useful in the analysis of gaseous samples [12]. The small volume and low gas flows required to sustain the glow discharge should provide low dead volume and minimize dilution effects for gaseous sample detection. The ability to sustain the glow discharge in helium offers the potential for determining higher ionization potential elements including Cl, As and Se. Interfacing GC with glow discharge is relatively straightforward and similar to our other experiences in interfacing GC with plasma techniques.

GC-GD should allow for excellent quantitation when the analytes of interest are introduced as gases. Initial studies have progressed well and are useful to report here [12]. The GC/GD interface developed is schematically indicated in Figure 4. The capillary column from the GC was connected to the glow discharge cell using a heated transfer line and the outlet of the column was centered in the sampling cone orifice. Reduced-pressure conditions in the GD cell were achieved using the expansion stage pumping of the mass spectrometer. The rf glow discharge was operated at 40 W power, 800 mL/min helium gas flow (6.4×10^{-1} mbar), with aluminum 6161 T6 as the sputtering target. The use of aluminum provides a clean background spectrum with no peaks appearing above *m/z* 50. The transfer line was maintained at a temperature of 325°C. Helium discharge gas was introduced surrounding the column and was also heated. For many elements, helium gas provides more favorable S/B than other gases such as argon and nitrogen [13]. Also, the uncompromised ability to work at the most abundant isotopes of 78 and 80 *m/z* for Se without interference from polyatomic ions [12,13] make it a reasonable support gas for these experiments.

With this interface and under the described conditions, the preliminary data are excellent and the results are presented in Figures 5 and 6 below. Figure 5 shows the chromatogram obtained for a 5 ng injection (each as Sn) of tetraethyltin (TET) and tetrabutyltin (TBT). Detection limits for both tin compounds (3 σ) were found to be 1 pg. The reproducibility for both compounds was less than 10% RSD. In these experiments, the GC was not solvent vented and therefore an extra contribution to baseline noise was present due to destabilization of the plasma. Our previous GC/ elemental analysis studies show that a vented experiment produces about ten-fold lower detection limits. With little more than this improvement the detection levels should equal or better the sub-pg levels we have achieved earlier by gaseous sample introduction to the MIP-MS (0.1 - 0.4 pg Sn) [8] or by SFC-ICP-MS (0.04 pg Sn) [7].

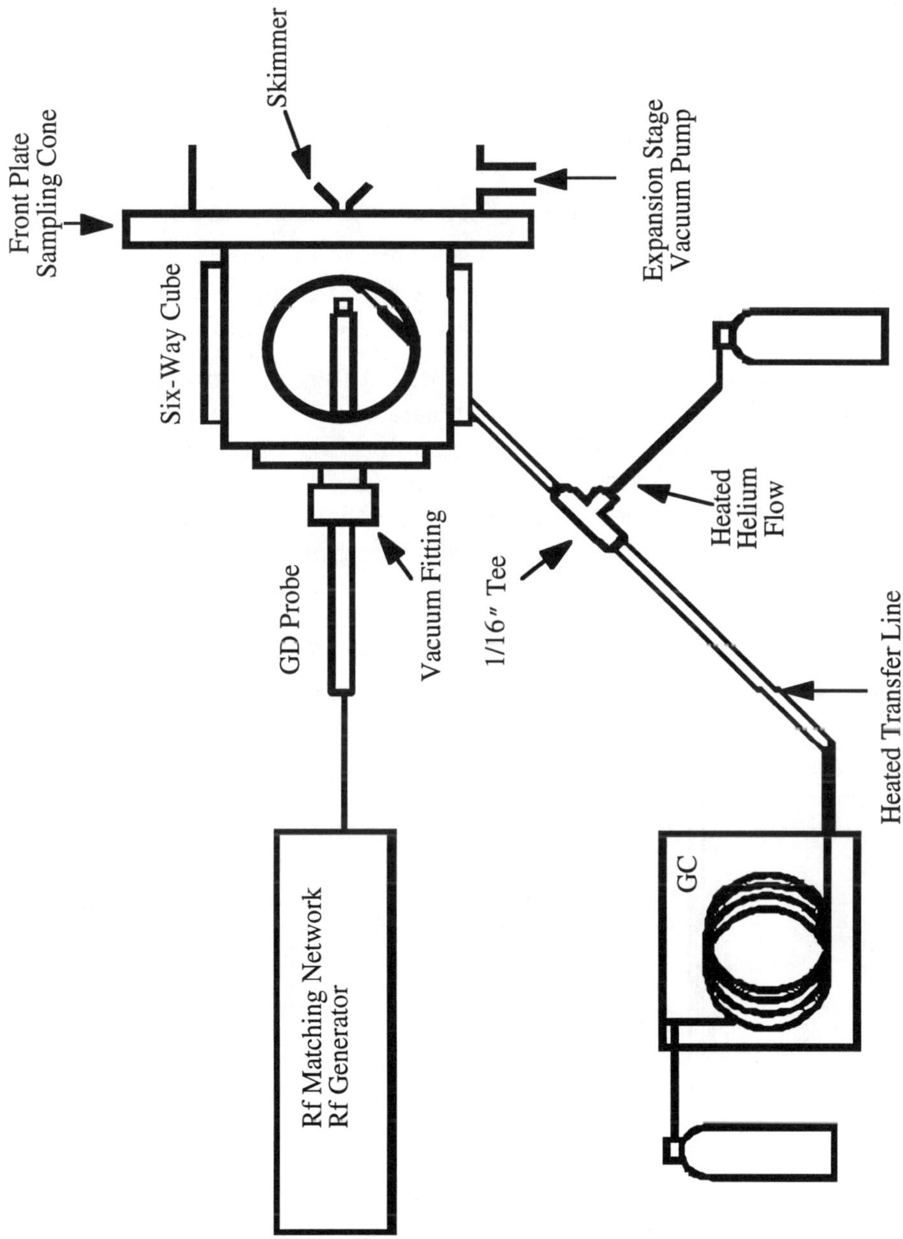

Figure 4. *GC - rf Glow Discharge Interface.*

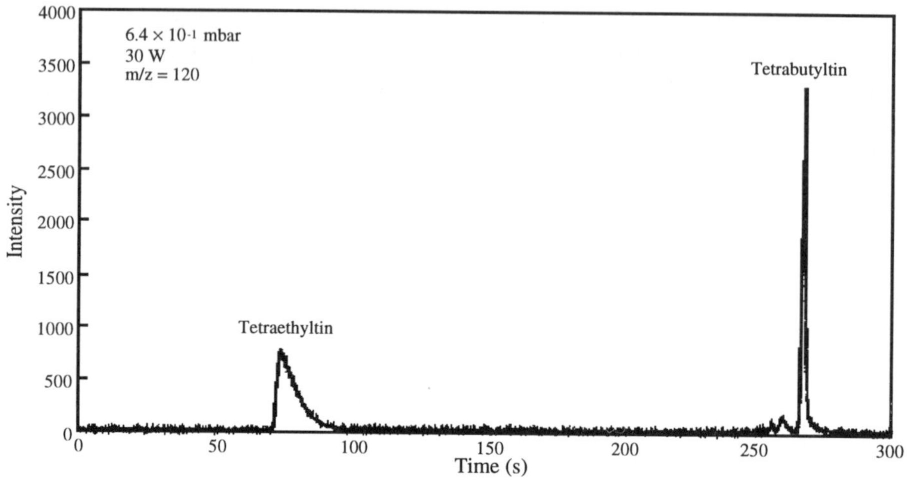

Figure 5. ***GC-GD Chromatogram for a 5 ng Injection of Each Organotin Compound.***

The potential for this GC-GD system to obtain structural information for these analytes is illustrated in Figure 6. This mass spectrum was obtained as the 5 ng tetraethyltin peak eluted. No changes in glow discharge operating conditions were necessary to obtain this spectrum. Note that both Figures 5 and 6 have intensity scales on the same order of magnitude and at approximately the same S/N, demonstrating in this way that it is possible to obtain useful structural information. This can be further refined by scanning to masses of potential interest such as those at 135 and 150 which correspond to monomethyltin and dimethyltin fragment ions, respectively.

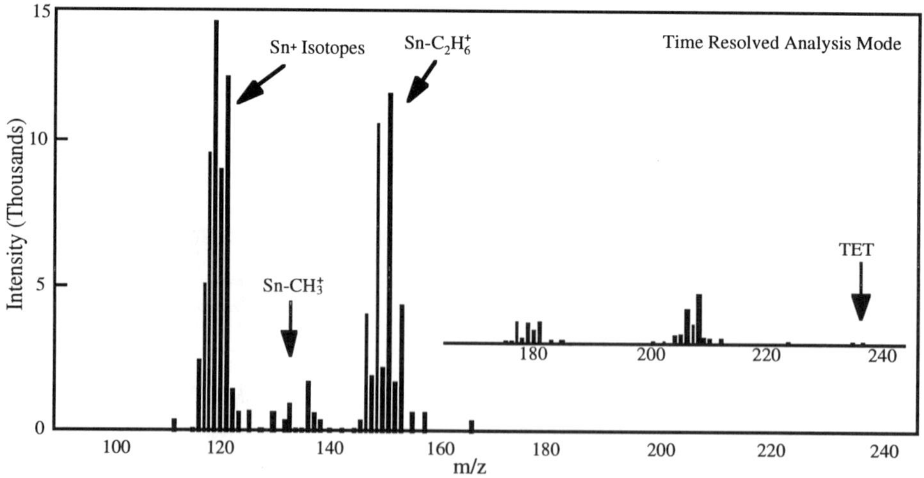

Figure 6. ***Mass Spectrum from Eluting Tetraethyltin $(C_2H_5)_4Sn$ Peak.***

Figure 7 illustrates the chromatograms obtained by single ion monitoring at m/z 150 and 235 for the same 5 ng injection of tetraethyltin and tetrabutyltin. This figure shows the potential for positive confirmation of identity of eluting peaks by looking at different m/z that might correspond to the bare element, to fragments or to the molecular ion. Such structural determinations with ultra-trace level elemental analyses are a new and important departure in elemental speciation studies. These preliminary experiments show high potential for elemental speciation studies to move beyond elemental species identification solely by retention time comparisons.

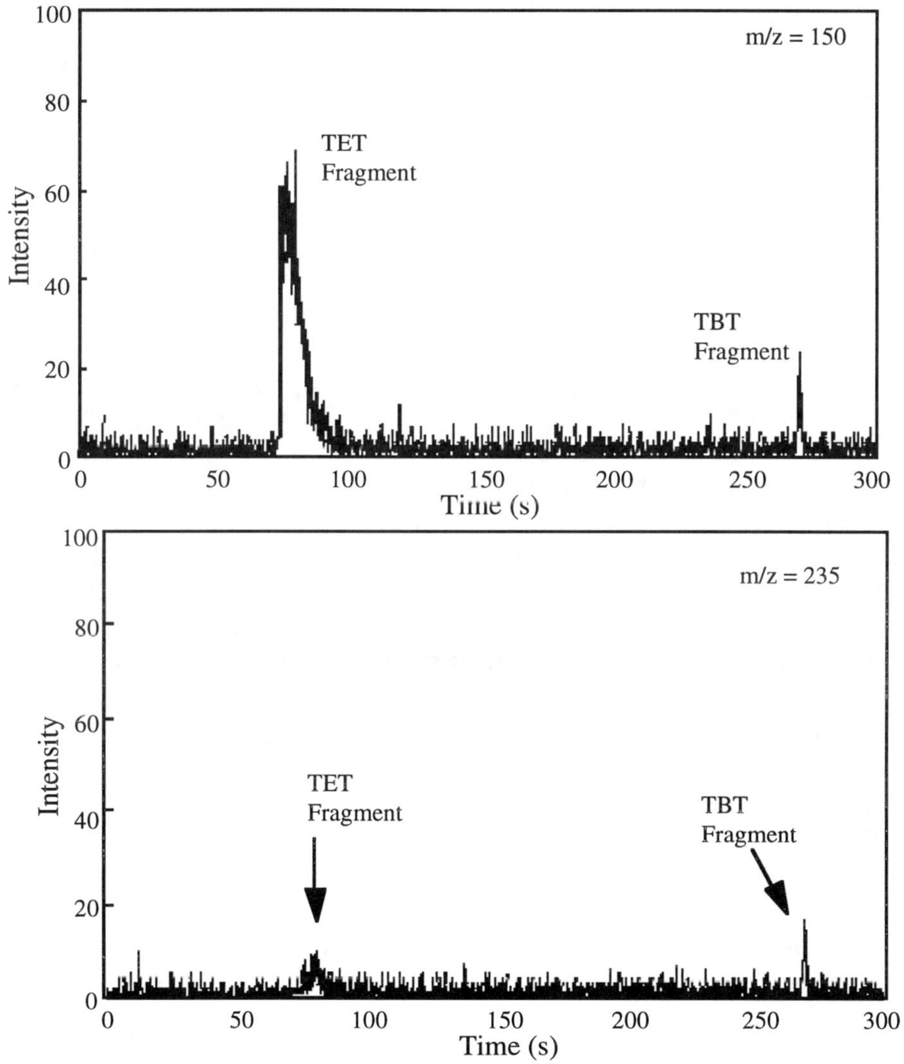

Figure 7. ***Single Ion Monitoring at m/z 150 and 235 for the Injection of a Mixture of Tetraethyltin and Tetrabutyltin (5 ng injection each).***

4. CAPILLARY ELECTROPHORESIS METHODS WITH ICP-MS DETECTION

CE is a relatively new separation technique that is potentially applicable to a wide range of polarity and molecular masses [14-16]. Hence, the potential for its application to speciation in complex samples of environmental health significance is high. Due to increasing sample complexity, higher chromatographic resolution is imperative. As a consequence, highly efficient methods that provide good separation in short times, such as with CE are required. However, since sample introduction volumes are very low, at 100 nL or less, detection levels for suitable CE detectors need to be improved. It is highly probable that plasma MS will provide enhanced detection capabilities for those species with metal functionality.

Major arsenic species reported in environmental and biological samples are arsenite, As (III); arsenate, As(V); dimethylarsinate, DMA; and monomethylarsonate, MMA. As a test for CE separation capabilities for inorganic arsenic and organoarsenicals, the above four arsenic species were determined using open tubular CE with indirect ultraviolet absorption detection [15]. Sodium chromate was used as a buffer solution and to provide the chromophore. The chromate absorption decreased as each As compound was sequenced past the detector. The absolute detection limits were 6.6, 4.4, 20, and 20 pg for arsenate, arsenite, MMA and DMA, respectively, with the concentration detection levels of 0.66, 0.44, 2.0 and 2.0 mg/L. A solution extracted at pH 10 from coal fly ash [6] was analyzed using this method and arsenate concentration was determined at 5 mg/L. Arsenite, MMA and DMA were not found. Quantitative results indicate that this method can be extended to the analysis of other environmentally significant sample types.

Two major challenges in CE-ICP-MS coupling are: 1) The small sample volume (10-100 nL) and low-flow rate (~ 2 μL/min with a 0.1 mm column) limit the choices for a nebulizer. The ultrasonic nebulizer (USN) is a reasonable choice for its ability to handle low sample flow rates, especially the μUSN. 2) High voltage is difficult to apply across the capillary since the end of the column is no longer placed into a beaker, as is the case in CE-UV. The ICP-MS must work as a post-column detector and therefore several interface designs were evaluated in our early work. Applying pressure post-column was unsuccessful, since even a simple separation was compromised. In this approach, moving from the "plug" flow of electrophoresis to laminar flow (pressurized) has a negative effect on the separation and as a consequence, the resolution for the separated peaks was lost.

Another CE-ICP-MS interface used the glass frit nebulizer. In this case, one end of the 50 μm i.d. fused silica CE column (60 cm length) was inserted through stainless steel tubing. Buffer solution passed over the capillary through the same stainless steel tubing into which a ground electrode also was inserted (the high voltage was applied at the other end of the column). This interface is shown in Figure 8.

Importantly, the electrical connection was continuous to the point of aerosol formation through the buffer solution. The analyte, together with the buffer solution, were introduced to the glass frit and the aerosol generated was introduced into the plasma. Two lead compounds, 8.5 ng injection of each $PbCl_2$ and $CH_3COO{\cdot}PbC_4H_9$, were easily separated using CE with ICP-MS detection [14]. It was found that the buffer solution flow rate affected the signal obtained using the glass frit nebulizer. Pulsing from the roller pump resulted in unstable background signal and this was eliminated by using an LC pump.

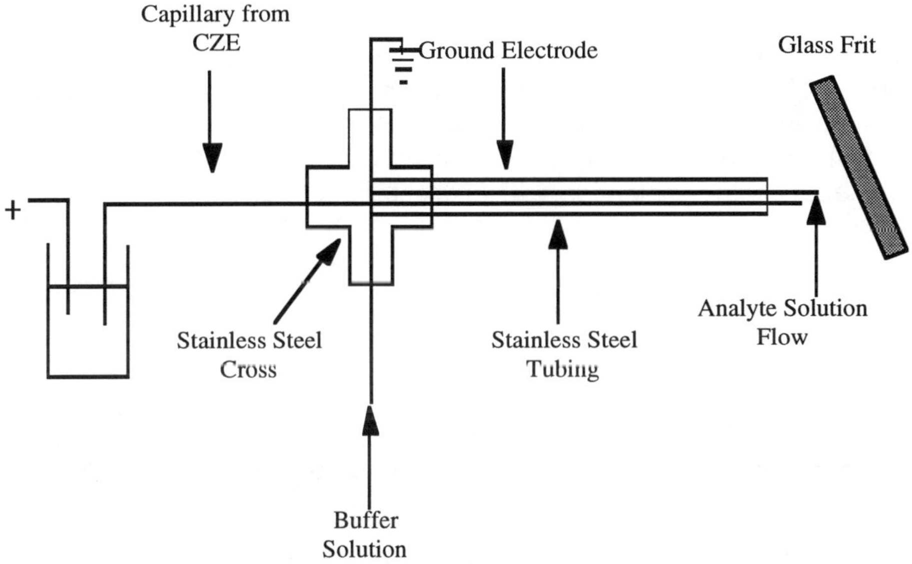

Figure 8. CE-ICP-MS interface design.

As we complete this initial study and include the four arsenic species, we will investigate USNs for aerosol generation. Better sensitivity, stability and detection limits are foreseen with the USN. Initial evaluations of CE-ICP-MS with a USN utilizing a 100 μL/min make-up buffer flow (for a circuit ground connection) around the CE capillary outlet have indicated that a stable electrophoretic current of 20 μA could be maintained when using applied voltages of 30 kV. This current was essentially the same as that when both ends of the CE capillary were immersed in buffer, indicating that this design was effective in maintaining the necessary ground connection. Further work is proposed to carry out electrophoretic separations of metal-based anions/cations using this interface. Improvements to the current USN system will include use of a ceramic tube instead of a stainless steel tube to reduce possible metal contamination. The use of a capillary tip coated with conducting paint to complete the ground connection will be tested with the USN design. This is similar in philosophy to that of Olesik *et al.* [16], who used metal paint used to coat the capillary of a concentric nebulizer. However, the USN is a different nebulizer type. Some buffer will likely be required for the USN to provide aerosol reproducibly.

5. CONCLUSIONS

Studies in elemental speciation continue growing and expanding in new areas in order to provide more information about the sample. Combination of different complementary techniques will allow for the separation, identification and quantitation of analytes present in complex matrices. New departures for elemental speciation studies using plasma mass spectrometric detection include the use of a low-power atmospheric pressure microwave induced plasma and GC-GD. These techniques show potential for elemental and structural studies at ultratrace levels.

REFERENCES

1. Vela, N.P., L.K. Olson and and J.A. Caruso, "Elemental Speciation with Plasma Mass Spectrometry," *Anal. Chem.* 65:585A-597A (1993).

2. Byrdy, F.A. and J.A. Caruso, "Elemental Analysis of Environmental Samples: Chromatography Coupled with ICP-MS Detection," *Environ. Sci. Technol.* 28:528A-534A (1994).

3. Byrdy, F.A. and J.A. Caruso, "Trace Metal Speciation by HPLC with Plasma Source MS Detection," *Environ. Health Perspect.* 103:21-23 (1995).

4. Byrdy, F.A. and J.A. Caruso, "Selective Chromatographic Detection by Plasma Mass Spectrometry," in *Selective Detectors for Chemical Analysis*, Vol. 131, Chapter 7, pp. 171-207. John Wiley & Sons (New York, NY 1995).

5. Tomlinson, M., J. Wangand and J.A. Caruso, "Speciation of Toxicologically Important Transition Metals Using Ion Chromatography with ICP-MS Detection," *J. Anal. At. Spectrom* 9:957-964 (1994).

6. Wang, J., M. Tomlinson and J.A. Caruso, "Extraction of Trace Elements in Coal Fly and Sequential Speciation by High Performance Liquid Chromatography with Inductively Coupled Plasma Mass Spectrometry (HPLC-ICP-MS)," presented at the 41st ASMS (American Society for Mass Spectrometry Conference), San Francisco, CA. June, 1993.

7. Vela, N.P. and J.A. Caruso, "Evaluation of Inductively Coupled Plasma Mass Spectrometry as an Elemental Detector for Supercritical Fluid Chromatography," *Anal. Chem.* 63:1491-1496 (1991).

8. Suyani, H., J. Creed, J.A. Caruso and R.D. Satger, "Helium Microwave-induced Plasma Mass Spectrometry for Capillary Gas Chromatographic Detection: Speciation of Organotin Compounds," *J. Anal. At. Spectrom.* 4:777-782 (1989).

9. Olson, L.K., W.C. Story, J. Creed, W-L. Shen and J.A. Caruso, "Fragmentation of Organic Compounds Using Low-Pressure Microwave Induced Plasma Mass Spectrometry," *J. Anal. At. Spectrom* 5:471-475 (1990).

10. Shen, W.L. and R.D. Satzger, "Development of a Low-Power Microwave Atmospheric Pressure Molecular Ionization Source for Mass Spectrometry with Direct Introduction of Gaseous and Liquid Organic Samples," *Anal. Chem.* 63:1960-1964 (1991).

11. Vela, N.P., D. Satzger, J. Giglio and J.A. Caruso, "An Atmospheric Pressure Low-Power Microwave Plasma Ionization Source for Liquid Chromatography - Mass Spectrometry (LC - MS)," presented at the 41st ASMS (American Society for Mass Spectrometry Conference), San Francisco, CA. June, 1993.

12. Caruso, J.A., L.K. Olson and J. Giglio, "Interfacing Gas Chromatograpy with Rf Glow Discharge Mass Spectrometry," 1994 FACSS, presentation # 645, St. Louis, MO.

13. Giglio, J.J., E.H. Evan and J.A. Caruso, "Evaluation of Alternative Plasma Gases for Radio-frequency Glow Discharge Mass Spectrometry (rf GD-MS)," *Appl. Spec.* 49:900-906 (1995).

14. Lin L., J.S. Wang and J.A. Caruso, "Capillary Zone Electrophoresis with Inductively Coupled Plasma Mass Spectrometry," Winter Conference on Plasma Spectrochemistry. January 1994, San Diego, CA.

15. Lin L., J.S. Wang and J.A. Caruso, "Arsenic Speciation Using Capillary Zone Electrophoresis (CZE) with Indirect UV Detection," *J. Chromatogr. Sci.* 33:177-180 (1995).

16. Olesik, J.W., J.A., Kinzer and S.V. Olesik, "Capillary Electrophoresis Inductively Coupled Plasma Spectrometry for Rapid Elemental Speciation," *Anal. Chem.*, 67:1-12 (1995).

METAL BINDING BY HUMIC SUBSTANCES IN SURFACE WATERS - EXPERIMENTAL AND MODELING CONSTRAINTS

E. Michael Perdue
School of Earth and Atmospheric Sciences
Georgia Institute of Technology
Atlanta, GA 30332

1. INTRODUCTION

Metal binding by humic substances is studied in both the field and laboratory. In field studies, experimental conditions are established by the chemical matrix of the natural sample, and metal binding studies generally consist of efforts to distinguish between free and complexed forms of a metal ion. Such a distinction is relevant because of the frequently strong correlation between biological effects of a metal ion and its free concentration [e.g., 1-3]. The major challenge in field studies is analytical methodology, i.e., making measurements of the total concentration of a trace metal ion, distinguishing between free and complexed forms of the metal ion, and conducting all phases of the study using clean techniques to avoid contamination of the sample. Recent advances in these areas are the topics of Chapters 6 and 7 in this volume.

In laboratory studies, metal binding can be studied over a range of experimental conditions that is hopefully extensive enough with respect to all experimental variables to encompass commonly encountered field conditions. Such laboratory studies often culminate in an effort to summarize the acquired knowledge of metal binding by humic substances through the development of chemical speciation models, the best of which consider competitive effects of H^+, other metal ions, and other ligands [e.g., 4-8]. The purpose of this paper is to examine the general relationship between experimental conditions, the extent of complexation of a trace metal by humic substances, and the characteristics and fitting parameters of chemical speciation models.

2. BACKGROUND

Humic substances are complex mixtures of organic compounds, whose complexity is so great that modern separation methods cannot resolve them into appreciable quantities of pure compounds. Accordingly, the study of any physical or chemical property of humic substances is, above all else, the study of a complex mixture. Unfortunately, chemists are generally trained to work on pure compounds and to avoid mixtures altogether. The following quote from a general chemistry text by Hammond *et al.* [9] clearly states the prevailing view of chemists:

> It is not always easy to determine whether a sample of material is a homogeneous mixture or a single substance. This is an important question to answer, for it is embarrassing for a chemist to find that he has been studying a mixture when he thought it was one substance.....The development of chemistry as a science and the systematic study of chemistry today depends on working with pure substances.

This "pure chemistry" paradigm must be discarded if unresolvably complex mixtures such as humic substances are to be studied in a scientifically acceptable manner. The first step in this direction is the realization that any measurable physical or chemical property of a humic substance is some kind of a weighted-average property, whose value must change as the composition of the mixture changes. The next step is to use experimental methods and mathematical models whose treatment of the weighted-average properties of a humic substance is as rigorous as possible.

In the scientific literature that pertains to humic substances, misleading phrases such as "the structure of the humic acid molecule...", "the molecular weight of fulvic acid...", "the conformation of the humic acid molecule...", or "the equilibrium constant for reaction of fulvic acid with..." are commonly encountered. The ubiquity of such phrases demonstrates that humic substances are often treated conceptually as pure substances, even by those who recognize that they are complex mixtures. Nowhere is this more true than in the study of metal binding by humic substances.

It is generally understood today that humic substances contain many non-identical metal binding sites. The experimental challenge is to measure the relative concentrations and strengths of those binding sites. As it turns out, the perceived distribution of binding sites is directly related to experimental conditions and to the type of conceptual model that is forced on the experimental data. These relationships will be examined more closely in this chapter.

3. COMPLEXATION CAPACITY

In the discussion of metal binding by humic substances, it is necessary to distinguish between actual and observable binding sites. Actual binding sites are chemical moieties that can form one or more bonds to a given metal ion, and the sum of their concentrations is the *complexation capacity* of the humic substance (often given in units of mmol/g). Observable binding sites are those binding sites whose metal content can be altered noticeably by manipulation of experimental conditions, e.g., by adding more metal, changing pH, etc. Clearly, observable binding sites are a variable subset of the actual binding sites.

The extent to which a binding site can be occupied by a given metal ion is a function of the nature and concentration of the metal ion, competitive reactions of H^+ and other cations with the binding site, competitive reactions of OH^- and other ligands with the metal, ionic strength, and the concentration of the humic substance [10,11]. The binding sites are structural entities that exist whether or not they are occupied by a metal ion under a specific set of experimental conditions, so the complexation capacity should be constant for a given mixture of binding sites.

The segregation of the existence of a binding site from the energetics of metal binding to that site is not only formally correct but also practically essential if reactions involving humic substances are to be incorporated into chemical speciation models. It is therefore worthwhile to discuss the relationship between experimental conditions and the apparent complexation capacity of a humic substance. It will be evident that experimental attempts to measure the complexation capacity of a humic substance must always yield results that are always less than the actual concentration of binding sites.

Consider a single binding site whose concentration is C_L mole/L and whose stability constant for reaction with a metal ion (M) is K. In the absence of competitive binding of other cations and ligands, the total concentrations of binding site and metal are:

$$C_L = [L] + [ML] \tag{1}$$

$$C_M = [M] + [ML] \tag{2}$$

and the degree of saturation (α) of the binding site with metal is given by:

$$\alpha = [ML]/C_L = (C_M - [M])/C_L = K[M]/(1+K[M]) \tag{3}$$

To actually saturate a binding site with metal, [M] must be increased sufficiently that α approaches 1.0 as closely as possible.

When an attempt is made to measure the complexation capacity of a humic substance, there are several experimental constraints on the values of [M], C_M and C_L that cause the complexation capacity to be underestimated. First of all, most metals that form strong complexes with humic substances also form sparingly soluble hydroxides. If the metal hydroxide is formed by the reaction:

$$M^{n+} + nH_2O - nH^+ = M(OH)_n(s),$$

then the maximum concentration of M^{n+} which can be used in the complexation capacity measurement is given by:

$$\log[M^{n+}] = -\log K_{MOH} - npH \tag{4}$$

For example, if the complexation capacity of a humic substance is measured at pH 7 using Cu^{2+}, the maximum concentration of Cu^{2+} that can be used is $10^{-5.3}$ mole/L. According to Equation 3, any binding site whose stability constant for binding copper (II) is less than $10^{+5.3}$ will be less than half-saturated with Cu^{2+} under these conditions. Obviously, the complexation capacity of the humic substance will be underestimated.

An obvious strategy for trying to overcome this limitation is to measure the complexation capacity at lower pH values. Unfortunately, competitive binding by H^+ (ignored in Equations 1-3) increases dramatically at lower pH values, so much so that the fraction of binding sites which can be occupied by copper(II) decreases at lower pH values. Thus, the complexation capacity will be even further underestimated.

Other precipitation reactions can also occur, possibly restricting even further the upper limit of [M] which can be used in the complexation capacity measurement. In water samples that are in equilibrium with atmospheric CO_2, for instance, metals can form carbonates and hydroxy carbonates. At sufficiently high concentrations, most metals cause humic substances to flocculate from solution. Once again, such restrictions on the upper limit for [M] inevitably cause the complexation capacity of the humic substance to be underestimated.

The last factor to be discussed here is experimental error in the measurements of C_M and [M]. An excellent analysis of the general effects of measurement errors on the estimates of binding site concentrations and stability constants was published a decade ago by Fish and Morel [12]. Because humic substances are a mixture, "complexed metal" is also a mixture - one that is not amenable to direct quantification. The concentration of complexed metal is therefore generally calculated as the difference between C_M and [M], corrected as required for inorganic complexes of the metal. Such calculations are most reliable when $C_M >> [M]$. If C_M and [M] are approximately equal, however, relatively small errors in C_M and [M] (especially the latter) can cause huge errors in the concentration of complexed metal, making it impossible to obtain reasonable estimates of complexation capacity.

To understand why C_M and [M] may be approximately equal under some experimental conditions, it is useful to return to the example of a single binding site that reacts with M. According to Equation 3, the degree of saturation of a binding site is determined by [M] and is independent of C_L. For example, a binding site with a stability constant of $10^{+4.0}$ is half-saturated at $[M] = 10^{-4.0}$ mole/L, regardless of the value of C_L. The relationship between C_M and [M], however, is quite sensitive to C_L. If C_L is $10^{-3.0}$ mole/L, then, at half-saturation of the binding site, $C_M = 6.0 \times 10^{-4}$ and [M] is 1.0×10^{-4}. A five percent error in [M] will only cause a one percent error in C_M - [M]. In contrast, if C_L is $10^{-6.0}$ mole/L, then $C_M = 1.005 \times 10^{-4}$ and [M] is 1.0×10^{-4}. In the latter case, a five percent error in [M] causes a 1000 percent error in C_M - [M], making it highly unlikely that the concentration of complexed metal can be calculated accurately.

More generally, the ratio of C_M/C_L which is required to achieve a particular degree of saturation of the binding site with metal is given by:

$$C_M/C_L = \alpha + (\alpha/(1-\alpha))/KC_L \qquad (5)$$

This relationship is presented graphically in Figure 1 for selected α values. If the product term KC_L is sufficiently large, only a small excess of metal is required to approach saturation ($\alpha = 0.99$) of a binding site with a metal. On the other hand, a large excess of metal is required for binding sites with lower KC_L values.

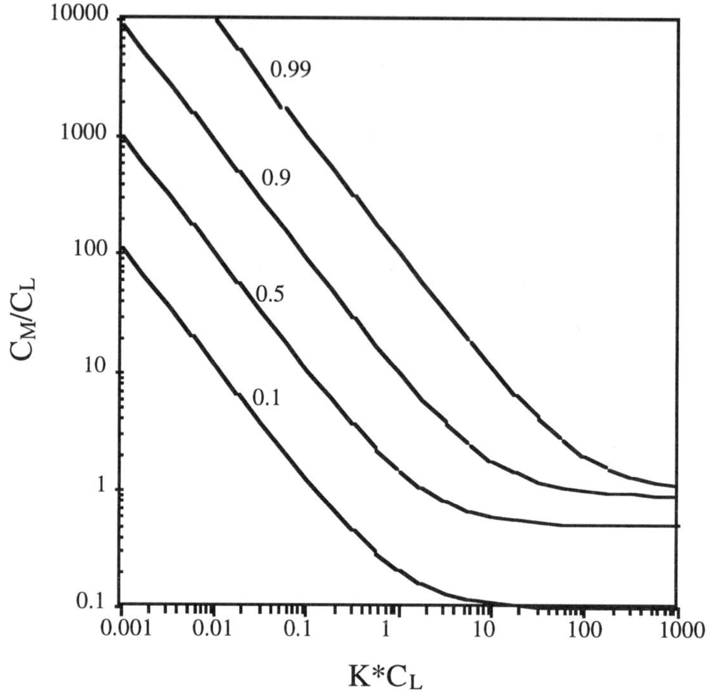

Figure 1. Total Metal: Total Ligand Ratios at Which Selected Degrees of Saturation of a Binding Site Can Bbe Achieved.

In practice, when an attempt is made to measure the complexation capacity of a humic substance, the error in C_M - [M] can become the factor which dictates the upper limit of [M] that is used. If that concentration is below the threshold for formation of metal-containing precipitates, then the upper limit for [M] can be increased simply by increasing C_L. This increases the product term KC_L for all binding sites in the humic substance, allowing a larger fraction of binding sites to be occupied with metal at a given C_M/C_L ratio. For this reason, attempts to measure complexation capacity in the laboratory at relatively high concentrations of a humic substance will always yield much higher estimates of complexation capacity than measurements made at natural environmental concentrations of the humic substance.

All the factors that have been considered cause the complexation capacity of a humic substance to be underestimated. The highest experimental estimates of complexation capacity that can be obtained will most likely be made using an experimental protocol that includes:

- neutral-to-alkaline pH
- low ionic strength
- high concentration of the humic substance

- strongly binding metal ion that does not form inorganic precipitates

The third of these conditions favors flocculation of the humic substance, and the last condition is simply hard to meet. Accordingly, most measurements of the complexation capacity of a humic substance are conducted under less-than-optimal conditions, and reported metal ion complexation capacities are certainly too low.

This author favors the use of the total acidity of a humic substance as an upper limit for its metal complexation capacity. For example, a humic substance that has a total acidity of 14 meq/g is assumed to have a metal complexation capacity of 14 meq/g (14.0, 7.0, and 4.7 mmol/g for monovalent, divalent, and trivalent cations, respectively). Such an approximation is consistent with the fact that any site which can bind H^+ must have a finite affinity for any other cation, and the use of equivalent, rather than molar, concentrations prevents the humic substance from acquiring a net positive charge through binding of polyvalent cations.

4. ENERGETICS OF METAL BINDING BY HUMIC SUBSTANCES

In most experimental studies of metal binding by humic substances, [M] is measured as C_M is varied over as wide a range as possible at constant pH (see previous section). From such data, the concentration of metal-humic complexes is easily calculated as C_M - [M]. Such experiments will be simulated in this paper, and titration curves will be presented in the form of α versus pM (-log[M], mole/L). In the interest of pedagogic clarity in subsequent calculations, all activity coefficients will be assumed to equal one (ideal behavior), and the formation of inorganic complexes will be neglected.

Humic substances are such a complex mixture that a chemically rigorous description of their reactions with a metal ion is unattainable. Accordingly, empirical models of many types have been used to describe the relationship between α and pM. Nearly all empirical models either implicitly or explicitly consider humic substances to contain a mixture of binding sites, each of which is characterized by a concentration (C_i/C_L) and a stability constant (K_i) for reaction with the metal ion. It may be possible to infer something about the actual distribution of binding sites in humic substances by considering the characteristics of models that have often been used to interpret the results of metal binding experiments. Comprehensive reviews of many common models are provided by Buffle [13], Sposito [14], and Perdue [15].

The modified Henderson-Hasselbalch equation, the modified Freundlich equation, and the Tanford electrostatic model are derived by empirical modifications of Equation 3, the equation for reaction of a metal ion with a single binding site. None of these models explicitly considers a distribution of binding sites; however, all three models are readily shown to be consistent with a broad symmetrical distribution of binding sites.

Consider first the modified Henderson-Hasselbalch equation, which has been used to "fix" a one-site model so that it can describe binding equilibria of large molecules containing many binding sites, e.g., natural and synthetic polymers. In its nonlinear form, this equation can be written as:

$$\alpha = K^{1/n}[M]^{1/n}/(1 + K^{1/n}[M]^{1/n}) \qquad (6)$$

where α is the mole fraction of binding sites that contain a metal ion (M), K is a binding constant, and n is a constant that is > = 1. Actually, if n equals one, Equation 6 describes the reaction of M with a single binding site (see Equation 3). The logarithm of this equation can be linearized in the following way:

$$pM = \log K - n \log (\alpha/(1-\alpha)) \qquad (7)$$

where pM is -log[M]. It is in this form that K and n were extracted from metal binding data prior to the widespread availability of computers. The distribution of binding sites that is implicit in this model can be observed in a plot of α versus pM, and even more clearly in a first derivative plot of -dα/dpM versus pM. The first derivative dα/dpM is given by:

$$d\alpha/dpM = -(\ln 10)\alpha(1-\alpha)/n \qquad (8)$$

Figure 2a and Figure 3a illustrate how the fitting parameter n affects plots of α versus pM and dα/dpM versus pM, respectively. In Figure 2a, the slope of the binding curve at α = 0.5 is greatest when *n* equals one and becomes progressively smaller as n is increased stepwise to three. The first derivatives of the curves in Figure 2a are plotted in Figure 3a. As n is increased from one to three, the curves become considerably broader, indicating that [M] is buffered over a wide pM range. Such buffering is only possible if there are binding sites whose log K values are distributed through that pM range.

The modified Freundlich equation originated in the study of the adsorption of gases on heterogeneous surfaces of catalysts [16], and this general equation has found broad applicability for modeling adsorption data in a variety of disciplines. It is usually used after the Langmuir equation (the adsorption equivalent of Equation 3) has been found to be unsatisfactory. In its nonlinear form, the modified Freundlich equation is:

$$\alpha = K[M]^{1/n}/(1 + K[M]^{1/n}) \tag{9}$$

By inspection, the modified Henderson-Hasselbalch and Freundlich equations are nearly identical. In fact, the K in the modified Freundlich equation is equivalent to the term $K^{1/n}$ in the modified Henderson-Hasselbalch equation, and n is the same in both equations. Not surprisingly, the logarithmic form of this equation:

$$pM = n \log K - n \log (\alpha/(1-\alpha)) \tag{10}$$

is nearly identical to the logarithmic form of the modified Henderson-Hasselbalch equation. The first derivative expression

$$d\alpha/dpM = -(\ln 10)\alpha(1-\alpha)/n \tag{11}$$

is, in fact, identical to the corresponding derivative in the modified Henderson-Hasselbalch equation (see Equation 8). The curves in Figure 2b and Figure 3b correspond exactly to the curves in Figure 2a and Figure 3a. Once again, the trends with increasing values of n are consistent with a broad distribution of stability constants.

The simple Tanford electrostatic equation, unlike the two previous models, is not an entirely empirical effort to fit data. It is reasonable to assume, on electrostatic grounds, that it should become progressively more difficult to add metal ions to a polyelectrolyte molecule as its binding sites become occupied with metal ions. This model postulates that, in the absence of electrostatic effects, all binding sites in the molecule have the same intrinsic K value, but that the "effective" K value decreases exponentially as sites are occupied with metal ions. In its nonlinear form, the simple Tanford electrostatic equation takes the form:

$$\alpha = K_0 \exp(Z\alpha)[M]/(1 + K_0 \exp(Z\alpha)[M]) \qquad (12)$$

where K_0 is the intrinsic binding constant and Z is a negative constant that determines how rapidly the "effective" K value changes as binding sites are filled with the metal ion. At $\alpha = 0.5$, it can be shown that the term ($\log K_0 + Z/(2 \ln 10)$) equals the K value in the modified Henderson-Hasselbalch equation. In logarithmic form, the simplified electrostatic equation becomes:

$$pM = \log K_0 - \log(\alpha/(1-\alpha)) + Z\alpha/\ln 10 \qquad (13)$$

and the corresponding first derivative expression is:

$$d\alpha/dpM = -(\ln 10)\alpha(1-\alpha)/(1 - Z\alpha(1-\alpha)) \qquad (14)$$

Figure 2c and Figure 3c illustrate the effect of Z on the binding curves and their first derivatives, respectively. Upon close inspection, it can be seen that the curves for the electrostatic model are somewhat different from those of the two preceding models, but not strikingly so. As Z is varied from zero (where Equation 12 is equivalent to Equation 3 for a single binding site) to - 8, the binding curves are less steep and the first derivative curves become much flatter. The increased buffering of [M] at high and low pM values is only possible if a humic substance contains binding sites with a broad distribution of stability constants for reaction with M.

One model that explicitly considers a broad distribution of binding sites is the Gaussian distribution model [17]. In that model, the probability of occurrence of a binding site in a humic substance is normally related to its log K value for reaction with M. For this model,

$$\alpha = \frac{1}{\sigma\sqrt{2\pi}} \int_{-\infty}^{+\infty} \left(\frac{K[M]}{1+K[M]}\right) \exp\left(-\left(\frac{\mu - \log K}{\sigma\sqrt{2}}\right)^2\right) d\log K \qquad (15)$$

where μ is the mean log K value for reaction with M and σ is the standard deviation of log K around the mean value (as σ approaches 0, Equation 15 approaches Equation 3 for one binding site). The corresponding first derivative equation is:

$$d\alpha / dpM = \frac{-(\ln 10)}{\sigma\sqrt{2\pi}} \int_{-\infty}^{+\infty} \left(\frac{K[M]}{(1+K[M])^2} \right) \exp\left(-\left(\frac{\mu - \log K}{\sigma\sqrt{2}} \right)^2 \right) d\log K \quad (16)$$

The binding curves and first derivative curves for the Gaussian distribution model are given in Figure 2d and Figure 3d, respectively. As σ is varied from zero to two, the distribution of binding sites changes from a single site to a broad continuous distribution. It is evident that increasing σ in the Gaussian distribution model, increasing n in the modified Henderson-Hasselbalch or modified Freundlich models, or decreasing Z in the Tanford electrostatic model has generally the same effect on binding curves and their first derivatives. The binding curves of humic substances cannot be fit by any of these equations unless $\sigma > 0$, $n > 1$, or $Z < 0$. It is logical to conclude that humic substances contain a broad distribution of metal binding sites.

It is worthwhile to consider further the general features of the binding curves in Figure 2 and their first derivatives in Figure 3. In Figure 2, all the models have a two-fold rotational axis of symmetry at $\alpha = 0.5$. Likewise, all the curves in Figure 3 have a plane of symmetry at the pM at which $\alpha = 0.5$ and also exhibit a maximum metal buffering intensity at that pM. These features are consistent with a symmetrical unimodal distribution of binding sites, such as the Gaussian distribution.

In contrast to the models which explicitly or implicitly consider a unimodal symmetrical distribution of binding sites, multi-site models are not constrained to be either unimodal or symmetric. In general, multi-site models have the form:

$$\alpha = \frac{[ML]}{C_L} = \frac{C_M - [M]}{C_L} = \frac{\sum_i [ML_i]}{C_L} = \sum_i \left(\frac{C_i}{C_L} \right) \left(\frac{K_i[M]}{1 + K_i[M]} \right) \quad (17)$$

When these models are used, it is common to assume the existence of a small number (2-3) of binding sites, each of which is characterized by its C_i and K_i values. The complexation capacity (C_L) is simply assumed to be the sum of all C_i values. The results of Fish and Morel [12] for copper binding in a 5 mg C/L solution of fulvic acid at pH 6 are typical: $C_1/C_L = 0.017$, $\log K_1 = 10.52$; $C_2/C_L = 0.126$, $\log K_2 = 8.28$; $C_3/C_L = 0.858$, $\log K_3 = 5.63$.

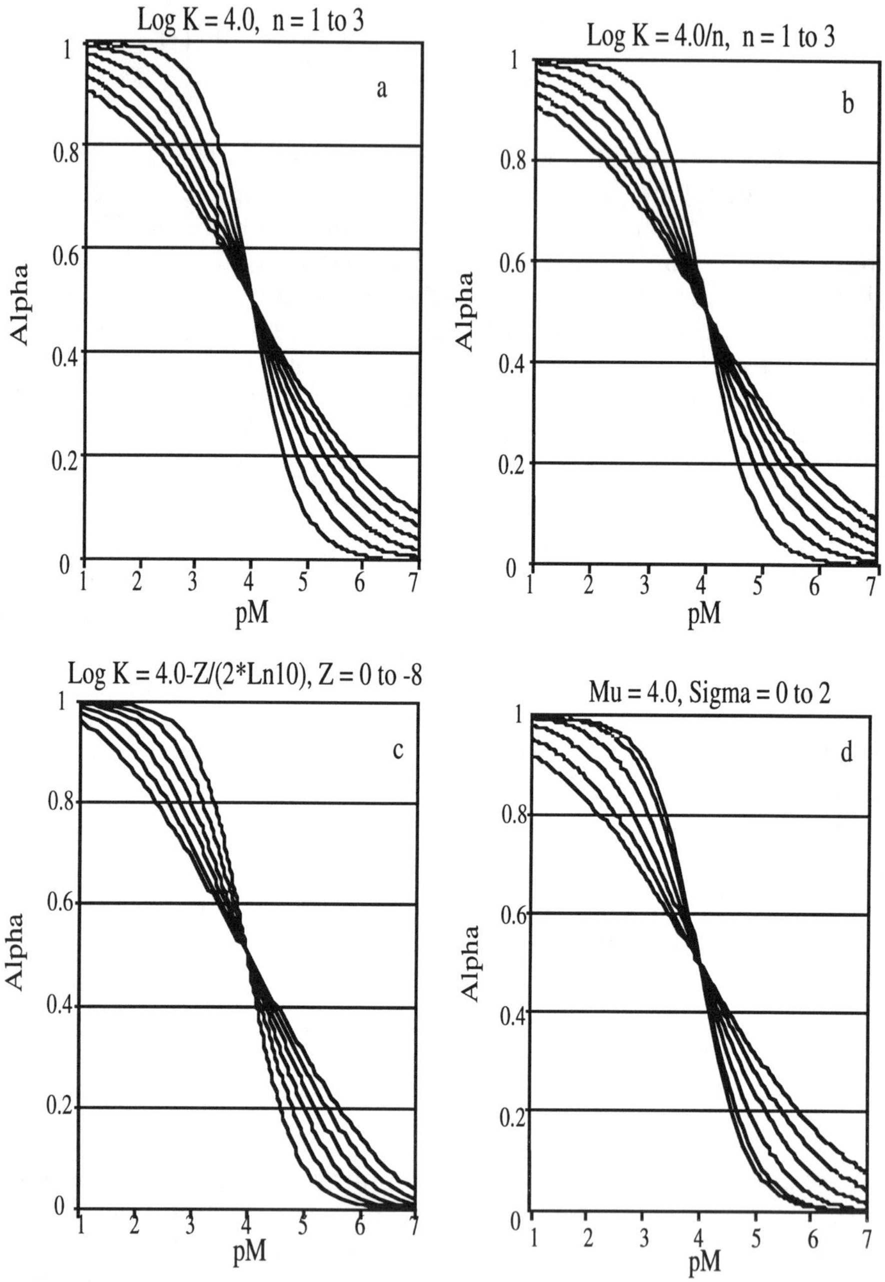

Figure 2. ***a) Henderson-Hasselbalch Equation, b) Freundlich Isotherm Equation, c) Electrostatic Equation, d) Gaussian Distribution Model.***

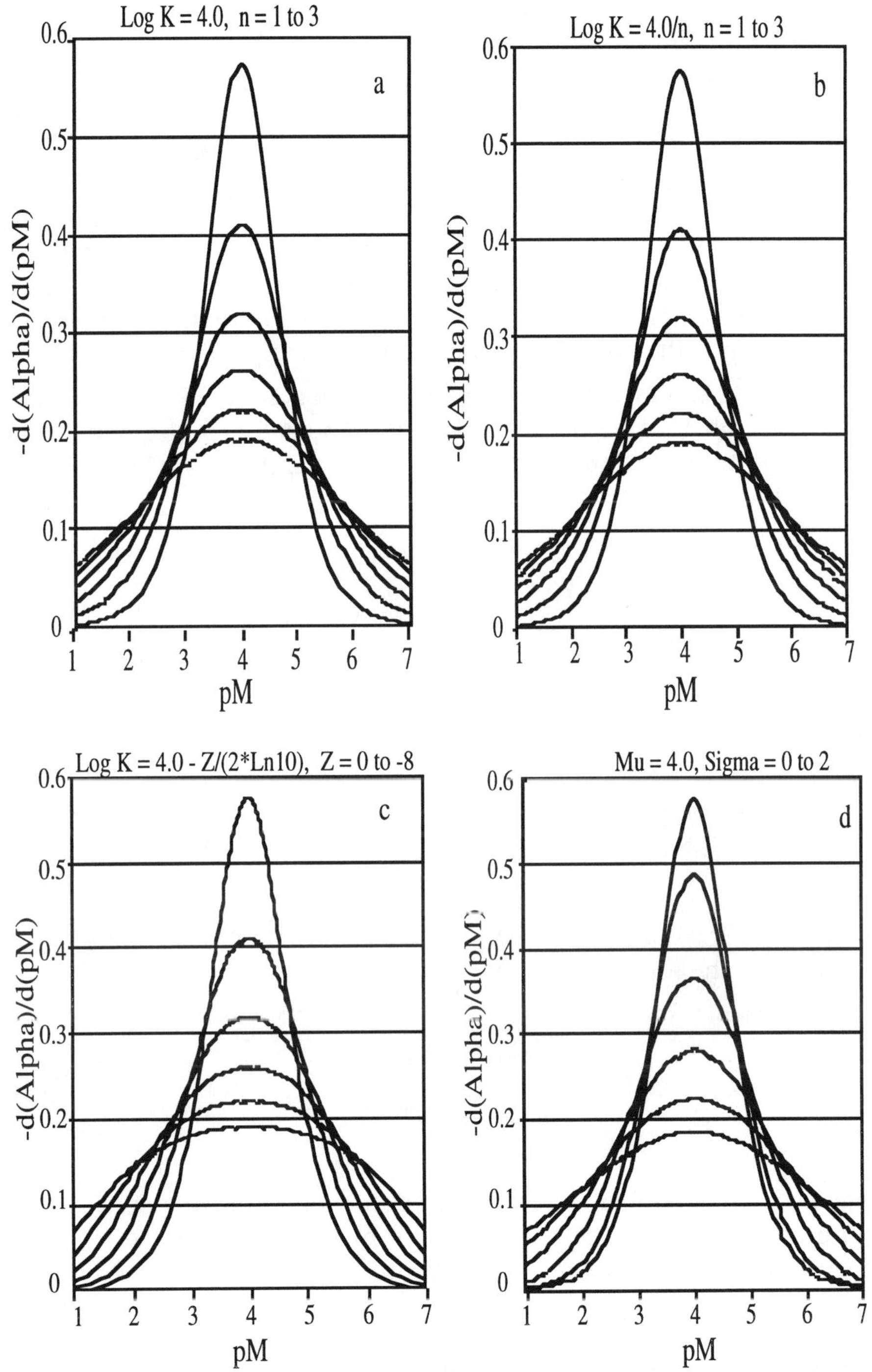

Figure 3. ***a) Henderson-Hasselbalch Equation, b) Freundlich Isotherm Equation, c) Electrostatic Equation, d) Gaussian Distribution Model.***

The general pattern of sharply decreasing C_i/C_L with increasing log K_i that was observed by Fish and Morel [12] is obtained whenever multi-site models are used to describe metal binding by humic substances. Perhaps not coincidentally, this pattern is also reminiscent of the high-log K side of a Gaussian distribution of binding sites. Given that there are no mathematical constraints on the fitting parameters of the three-site model, it is interesting to ask whether or not any relationship exists between those parameters and the fitting parameters of a Gaussian distribution model. To address this question, it is necessary to more closely examine the Gaussian distribution model and, particularly, how the model is evaluated numerically by this author.

In practice, the necessary numerical integration in the Gaussian distribution model (Equation 15,16) is efficiently performed using Gauss-Hermite quadrature [7,18]. For example, the integral in Equation 15 can be transformed into:

$$\alpha = \frac{1}{\sqrt{\pi}} \int_{-\infty}^{+\infty} \left(\frac{10^{\mu + Z\sigma\sqrt{2} - pM}}{1 + 10^{\mu + Z\sigma\sqrt{2} - pM}} \right) \exp\left(-Z^2\right) dZ \qquad (18)$$

where

$$Z = \left(\frac{\mu - \log K}{\sigma\sqrt{2}} \right) \qquad (19)$$

The integral can then be approximated quite accurately by a summation of the form:

$$\alpha = \frac{1}{\sqrt{\pi}} \sum_{i=1}^{N} W_i \left(\frac{10^{\mu + Z\sigma\sqrt{2} - pM}}{1 + 10^{\mu + Z\sigma\sqrt{2} - pM}} \right) \qquad (20)$$

where N is the number of terms used to approximate the Gaussian integral and Z_i and W_i are the Gauss-Hermite quadrature set of points and weights corresponding to a particular value of N. It is readily shown that Equation 20 is identical to the multi-site binding model (Equation 17), with C_i/C_L being equivalent to $W_i/\sqrt{\pi}$ and the remaining terms in the two equations being identical. In a proper set of Gauss-Hermite quadrature points and weights,

$$\sum_{i=1}^{N} Z_i^2 = N(N-1)/2 \quad \text{and} \quad \sum_{i=1}^{N} W_i = \sqrt{\pi} \tag{21}$$

Several sets of Gauss-Hermite quadrature points and weights for small values of N are given in Table 1.

Table 1. Points and weights for N-Point Gauss-Hermite Quadrature.

N	Pt./Wt.	i=1	i=2	i=3	i=4	i=5	i=6	i=7
3	Pt.	1.22474	0.00000	-1.22474				
	Wt.	0.29541	1.18164	0.29541				
4	Pt.	1.65068	0.52465	-0.52465	-1.65068			
	Wt.	0.08131	0.80491	0.80491	0.08131			
5	Pt.	2.02018	0.95857	0.00000	-0.95857	-2.02018		
	Wt.	0.01995	0.39362	0.94531	0.39362	0.01995		
6	Pt.	2.35060	1.33585	0.43608	-0.43608	-1.33585	-2.35060	
	Wt.	0.00453	0.15707	0.72463	0.72463	0.15707	0.00453	
7	Pt.	2.65196	1.67355	0.81629	0.00000	-0.81629	-1.67355	-2.65196
	Wt.	0.00097	0.05451	0.42561	0.81026	0.42561	0.05451	0.00097

It is interesting to compare the fitting parameters of the three-site model [12] with a subset of Gauss-Hermite points and weights from Table 1. It was argued previously that less than half of the binding sites in a humic substance can be titrated in a typical experiment, so the first three points and weights for N = 7 will be used. First of all, from Equation 19, the log K_i value corresponding exactly to Z_i is given by: $\log K_i = \mu + (\sigma\sqrt{2})Z_i$. A plot of log K_i versus Z_i should be linear, its intercept is μ, and σ can be calculated from the slope. In Figure 4a, the log K values from the three-site model [12] are plotted versus the first three Gauss-Hermite points for N = 7 in Table 1. Clearly, the agreement is excellent. The line in Figure 4a is a linear regression fit, for which $r^2 > 0.99$.

From Equation 21 and the symmetrical nature of the Gauss-Hermite weights, it is easily shown that the sum of the first three Gauss-Hermite weights for N = 7 must equal $(\sqrt{\pi} - W_4)/2$. It follows that the corresponding C_i/C_L should equal $2.08W_i$. This prediction is compared with the actual C_i/C_L values of the three-site model in Figure 4b. In this case, the theoretical line is plotted instead of a regression line. The agreement between the actual C_i/C_L values from the three-site model and the predictions for the first three Gauss-Hermite weights is excellent.

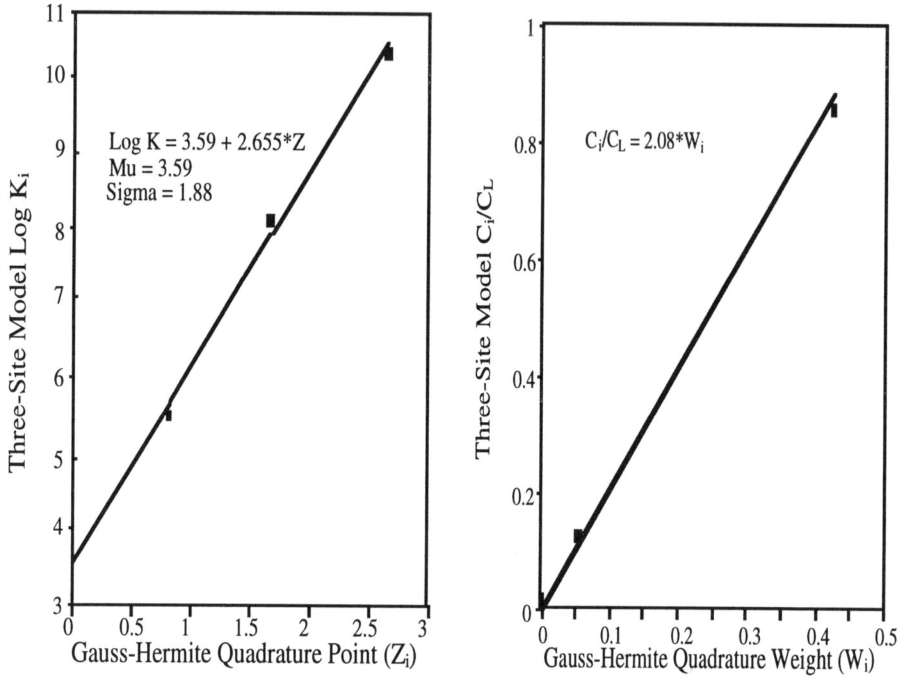

Figure 4. ***a) A Comparison of Log K_i Values from a Three-site Discrete Model with Gauss-Hermite Points, b) A Comparison of Log C_i/C_L Values from a Three-site Discrete Model with Gauss-Hermite Weights.***

In Figure 4a and Figure 4b, the log K_i and C_i/C_L values from the three-site model used by Fish and Morel [12] correlate strongly with the first three Gauss-Hermite points and weights in the set of points and weights for N = 7. This analysis doesn't really prove anything about the actual distribution of binding sites in the humic substance used by Fish and Morel [12], but it does show that their observations are fully consistent with a Gaussian distribution of binding sites.

5. FITTING MODELS TO DATA

In the preceding sections of the paper, the extent to which M is complexed by humic substances was expressed in terms of α, the degree of saturation of binding sites. From Equation 3, it is evident that α is calculated from experimental data as $(C_M - [M])/C_L$. Even if C_M and [M] are known, α cannot be calculated unless the complexation capacity (C_L) is known. As discussed previously, experimental estimates of C_L are invariably too low, and their use will cause α values to be overestimated. It is better to express the extent of metal binding in terms of the bound metal concentration [ML], which is simply C_M-[M]. All the preceding models can be transformed to use [ML] as the dependent variable, because [ML] is simply αC_L. Of course, C_L then becomes another fitting parameter in these models. As an example, the transformed Gaussian distribution model is:

$$[ML] = \frac{C_L}{\sigma\sqrt{2\pi}} \int_{-\infty}^{+\infty} \left(\frac{K[M]}{1+K[M]}\right) \exp\left(-\left(\frac{\mu - \log K}{\sigma\sqrt{2}}\right)^2\right) d\log K \qquad (22)$$

All the models in Figure 2 predict binding curves that have an inflection point at $\alpha = 0.5$. Actually, though, this author is not aware of any published data set in which an inflection point is evident. Accordingly, all efforts to model the distribution of metal binding sites in humic substances use data from the high pM side of Figures 2 and 3. It is possible that the weaker binding sites which might react at low pM do not exist at all, being instead a mathematical consequence of fitting one of the preceding models to experimental data. Alternatively, the experimental upper limits for [M], which were discussed previously, may be too low to half-saturate, much less saturate, the mixture of metal binding sites in a humic substance.

With the exception of the multi-site model, each of the preceding models explicitly or implicitly postulates the existence of a symmetrical distribution of binding sites. If C_L is determined by fitting [ML] as a function of [M], then the fitting parameters obtained from data collected at relatively high pM will reflect the existence of such a distribution. The fitted C_L value will probably be more than a factor of two greater than the highest measured [ML]. If the modified Henderson-Hasselbalch equation, the modified Freundlich equation, or the Gaussian distribution model is used, then log K, n(log K), or μ, respectively, will probably be smaller than the lowest experimental pM value. If the Tanford electrostatic equation is used, log K_0 will be greater than the pM at which the binding sites are half-saturated. If less than half of the sites can be saturated with M, however, then log K_0 may be smaller than the lowest experimental pM value.

When multi-site models are fit to experimental data, virtually any kind of distribution of binding sites is theoretically possible, because those models are not constrained to be either unimodal or symmetric. Even so, when experimental data are collected at relatively high pM, the

C_i/C_L and K_i values generally resemble those obtained by Fish and Morel [12]. There are two fitting parameters (C_i, K_i) that must be determined for each binding site in a multi-site model (see Equation 17), so a two-site or three-site model has enough adjustable parameters to fit typical experimental data sets.

When a multi-site model is used, the log K_i values that are obtained generally lie between the highest and lowest experimental pM values (pM_{high} and pM_{low}, respectively). For example, Fish and Morel [12] varied pCu between about 10.6 and 4.2 in the titration of a 5.0 mg C/L fulvic acid solution at pH 6.0, and their three-site model yielded log K_i values of 10.52, 8.28, and 5.63.

To better understand why multi-site models generally yield log K_i values that lie between pM_{high} and pM_{low}, it is useful to express $\Sigma[ML_i]$ from Equation 17 using three summations.

$$\sum_{i=1}^{N}[ML]_i = \sum_{i=1}^{j} C_i\left(\frac{K_i[M]}{1+K_i[M]}\right) + \sum_{i=j+1}^{k} C_i\left(\frac{K_i[M]}{1+K_i[M]}\right) + \sum_{i=k+1}^{N} C_i\left(\frac{K_i[M]}{1+K_i[M]}\right) \quad (23)$$

The first summation includes those binding sites for which $K_i[M] >> 1$ over the entire experimental pM range, the third summation includes those binding sites for which $K_i[M] << 1$ over the entire experimental pM range, and all other binding sites are included in the second summation. Accordingly, Equation 23 can be simplified to obtain:

$$\sum_{i=1}^{N}[ML]_i = \sum_{i=1}^{j} C_i + \sum_{i=j+1}^{k} C_i\left(\frac{K_i[M]}{1+K_i[M]}\right) + \sum_{i=k+1}^{N} C_i K_i[M] \quad (24)$$

This equation provides a good logical framework for understanding the relationship between log K_i values from a multi-site model and the experimental pM range. As [M] is varied from pM_{high} to pM_{low}, the three summation terms in Equation 24 exhibit quite different behavior.

The first summation term in Equation 24 is constant over the entire experimental pM range, and it equals the total concentration of very strong binding sites in the humic substance. If a multi-site model is fit to experimental data, all the very strong binding sites will seem to be a single site. The C_i of that "site" is actually the total concentration of the very strong binding sites, and the K_i value for that "site" is a large, meaningless number (large enough that $K_i[M] >>1$ for all [M]). It is certain, however, that the actual log K_i values of the very strong binding sites are substantially greater than pM_{high}.

The third summation includes those binding sites whose log K_i values are substantially less than pM_{low}, i.e., the degree of saturation of these binding sites with M is very low. It is these sites, in particular, that cannot be saturated with M in a typical complexation capacity measurement. The third summation term varies linearly with [M], with a slope of $(\Sigma_{i=k+1}^{N} C_iK_i)$. If a multi-site model is fit to experimental data for which $\Sigma[ML_i]$ varies linearly with [M], the resulting C_i and K_i values will be non-unique, i.e., any combination of C_i and K_i values that has the same $(\Sigma_{i=k+1}^{N} C_iK_i)$ will fit the data. Accordingly, neither C_i nor K_i values for any of the binding sites in this subset of binding sites can be obtained from experimental data.

All remaining binding sites are included in the second summation term of Equation 24. The log K_i values for this subset of binding sites range from slightly greater than pM_{high} to slightly less than pM_{low}. As M] is varied from pM_{high} to pM_{low}, the degree of saturation of these binding sites varies over a considerable range. This makes it possible to obtain both C_i and K_i values by using nonlinear regression to fit the multi-site model to experimental data. The resulting log K_i values generally lie within the experimental pM range, although it is possible to obtain log K_i values that lie slightly outside that range.

6. CONCLUSIONS

The extent to which the concentrations and strengths of metal binding sites in humic substances can be known is limited by experimental constraints in the measurements of C_M and [M] and by the properties of models that are fit to experimental data. At a given pH and C_L, [M] can only be increased until a metal hydroxide or carbonate precipitates, the humic substance is flocculated by reactions with M, or errors in measurements of C_M and [M] become so large that [ML] cannot be estimated reliably. Assuming that all proton binding sites have a finite affinity for other cations and that one mole of M^{n+} displaces n moles of H^+, the upper limit of [M] is generally reached before even half of the binding sites in a humic substance have reacted. Consequently, the complexation capacity of a humic substance is generally underestimated, and the fitting parameters of mathematical models of metal binding by humic substances are derived from experimental data for reactions of the relatively strong binding sites. All models discussed here except the multi-site model either explicitly or implicitly assume the existence of a continuous, unimodal, symmetrical distribution of log K_i values. The fitting parameters of those models reflect this underlying assumption. When a multi-site model is used to fit experimental data, it is likely that the set of log K_i values will lie within the experimental pM range. More surprisingly, the C_i and K_i values that were obtained using a three-site model correlated strongly with the first three Gauss-Hermite points and weights (from a set of seven) that are used to integrate a normal distribution of metal binding sites.

REFERENCES

1. Sunda, W. and R.R.L. Guillard, "The Relationship Between Cupric Ion Activity and the Toxicity of Copper to Phytoplankton," *J. Mar. Res.* 34:511-529 (1976).

2. Morel, F.M.M. and J.G. Hering, *Principles and Applications of Aquatic Chemistry*, John Wiley and Sons, pp. 405-414. (New York, NY 1993).

3. Erickson. R.J., D.A. Benoit, V.R. Mattson, H.P. Nelson, Jr. and E.N. Leonard, "The Effects of Water Chemistry on the Toxicity of Copper to Fathead Minnows," *Environ. Toxicol. Chem.* 15:181-193 (1996).

4. Ephraim, J. and J.A. Marinsky, "A Unified Physicochemical Description of the Protonation and Metal Ion Complexation Equilibria of Natural Organic Acids (Humic and Fulvic Acids). 3. Influence of Polyelectrolyte Properties and Functional Heterogeneity on the Copper Ion Binding Equilibria in an Armadale Bh Fulvic Acid Sample," *Environ. Sci. Technol.* 20:367-376 (1986).

5. Cabaniss, S.E. and M.S. Shuman, "Copper Binding by Dissolved Organic Matter: I. Suwannee River Fulvic Acid Equilibria," *Geochim. Cosmochim. Acta* 52:185-193 (1988).

6. Susetyo, W., J.C. Dobbs, L.A. Carreira, L.V. Azarraga and D.M. Grimm, "Development of a Statistical Model for Metal-Humic Interactions," *J. Anal. Chem.* 62:1215-1221 (1990).

7. Allison, J.A. and E.M. Perdue, "Modeling Metal-Humic Interactions with MINTEQA2," in *Humic Substances in the Global Environment and Implications on Human Health*, N. Senesi and T.M. Miano, Eds., pp. 927-942. Elsevier (Amsterdam, Holland 1994).

8. Tipping, E., "WHAM - A Chemical Equilibrium Model and Computer Code for Waters, Sediments, and Soils Incorporating a Discrete Site/Electrostatic Model of Ion-Binding by Humic Substances," *Computers Geosci.* 20:973-1023 (1994).

9. Hammond, G.S., J. Osteryoung, T.H. Crawford and H.B. Gray, *Models in Chemical Science - An Introduction to General Chemistry*, 422 pgs. Benjamin (New York, NY 1971).

10. Perdue, E.M., "Measurements of Binding Site Concentrations in Humic Substances," in *Metal Speciation - Theory, Analysis, and Application*, J.R. Kramer and H.E. Allen, Eds., pp. 135-154. Lewis Publishers (Boca Raton, FL 1988).

11. Perdue, E.M., "Effects of Humic Substances on Metal Speciation," in *Aquatic Humic Substances: Influence on Fate and Treatment of Pollutants*, I.H. Suffet and P. MacCarthy, Eds., pp. 281-295. American Chemical Society Advances in Chemistry Series No. 219 (Washington, DC 1989).

12. Fish, W. and F.M.M. Morel, "Propagation of Error in Fulvic Acid Titration Data: A Comparison of Three Analytical Methods," *Can. J. Chem.* 63:1185-1193 (1985).

13. Buffle, J., "Natural Organic Matter and Metal-Organic Interactions in Aquatic Systems,' in *Metal Ions in Biological Systems*, H. Sigel, Ed., pp. 165-221. Marcel Dekker (New York, NY 1984).

14. Sposito, G., "Sorption of Trace Metals by Humic Materials in Soils and Natural Waters," *CRC Critical Rev. Environ. Control* 16:193-229 (1986).

15. Perdue, E.M., "Modeling the Acid-Base Chemistry of Organic Acids in Laboratory Experiments and in Freshwaters," in *Organic Acids in Aquatic Ecosystems*, E.M. Perdue and E.T. Gjessing, Eds., pp. 111-126. John Wiley and Sons Ltd. (Chichester, UK 1990).

16. Sips, R., "On the Structure of a Catalyst Surface," *J. Chem. Phys.* 16:490-495 (1948).

17. Perdue, E.M. and C.R. Lytle, "Distribution Model for Binding of Protons and Metal Ions by Humic Substances," *Environ. Sci. Technol.* 17:654-660 (1983).

18. Parrish, R.S. and E.M. Perdue, "Computational Methods for Fitting Statistical Distribution Models of Multi-Site Binding Equilibria," *Math. Geol.* 21:199-219 (1989).

10

MODELING METAL SPECIATION IN AQUATIC SYSTEMS

John C. Westall[1], *Miroslav Černík*[2] *and Michal Borkovec*[2]
[1]Department of Chemistry
Oregon State University
Corvallis, OR 97331
[2]Institute for Terrestrial Ecology
Federal Institute of Technology (ETH)
Zurich, Switzerland

1. INTRODUCTION

Models are important for estimation of speciation of metals in surface waters, groundwaters, soils, and sediments. Speciation affects not only transport and fate of metals, but also their biological availability and environmental effects. Speciation models are thus needed in virtually all aspects of managing metal contamination of the environment, including risk assessment, site remediation, and waste disposal management.

Although metal speciation models based on chemical equilibrium have been applied in the environmental sciences for at least twenty-five years [1,2], the number of cases is still very small in which either (i) the output of the model has been "validated" by field measurements or (ii) the output of the model has actually been used as the primary basis for a decision in environmental management [3-5]. One of the greatest obstacles to the use of equilibrium models to predict metal speciation in field systems is the *heterogeneity of environmental sorbents*, such as humic substances and the surfaces of rocks and minerals. (Another major source of uncertainty, the slow rates of chemical reactions, will not be addressed in this paper.) Thus, while one can generally have some confidence in the database determined under laboratory conditions and the computer program used to compute speciation, it is the application to heterogeneous materials from the field that is veiled in uncertainty, and the reason that speciation models are used primarily as "a way of thinking about a problem" rather than as "the direct basis for managing metals in the environment."

In this paper we shall review the development of models for the binding of metals to heterogeneous environmental materials and describe some promising new approaches that should advance our ability to use metal speciation models successfully in field situations.

2. BACKGROUND

2.1 The Generic Metal Speciation Problem

The generic metal speciation problem for surface waters or groundwaters can be formulated in the following way (considering the solution phase, only). A metal ion, M^{n+}, forms complexes with simple inorganic ligands (e.g., OH^-, CO_3^{2-}, Cl^-, etc.), simple organic ligands of natural or synthetic origin (e.g., oxalate, NTA, etc.), and complex, heterogeneous environmental complexing agents (e.g., surfaces of sand grains, clay particles, humic substances, etc.).

If only the simple ligands were involved, field speciation problems could be solved relatively well with existing models, which already include the data necessary for these simple ligands. However, the presence of heterogeneous complexing agents greatly complicates the situation; these materials generally cannot be represented by just one or two reactive sites, particularly if the experimental data span a broad range.

There are of course many ways to incorporate heterogeneity in the model. In this paper we shall focus on a representation of the heterogeneous material as a collection of sites X_i, which react with a metal ion M according to the reaction (for 1:1 stoichiometry):

$$M + X_i = MX_i \qquad K_{MX_i} \qquad (1)$$

where ionic charges have been omitted for simplicity. The problem of characterizing the heterogeneous material then becomes one of determining the abundances of the sites X_i, the stability constants K_{MX_i}, and the reaction stoichiometries.

2.2 Approaches to Representing Heterogeneous Environmental Complexing Agents

Basically two approaches can be taken towards the characterization of these distributions of site abundances, stability constants, and reaction stoichiometries associated with the metal complexing behavior of heterogeneous environmental complexing agents, as outlined in Figure 1.

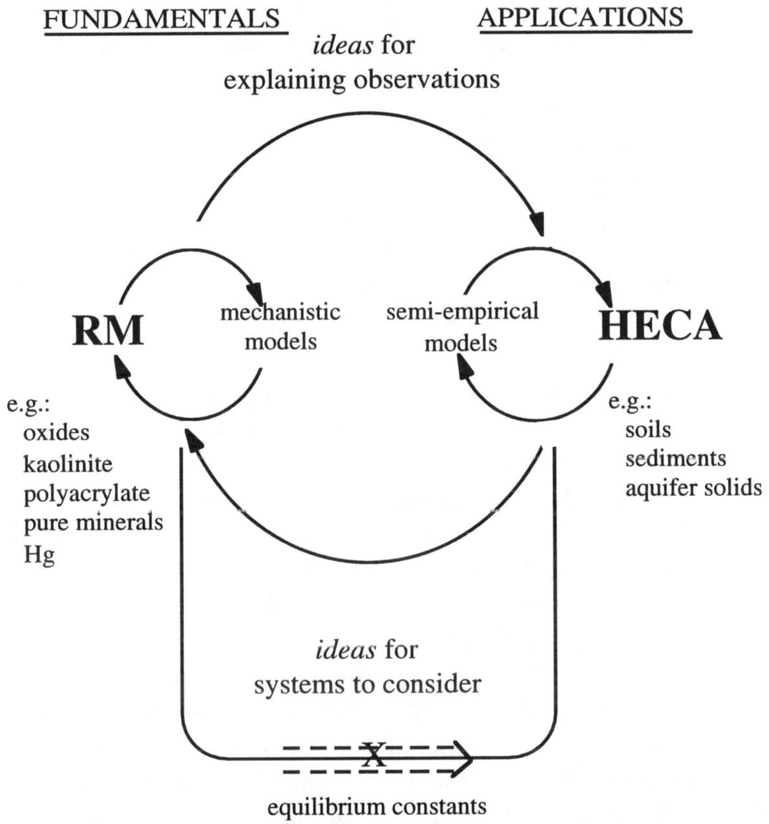

Figure 1. ***Representation of Iterative Approach to Environmental Research: on the Left-hand Side are Represented "Clean" Experimental Systems for Which Detailed Mechanistic Models May Be Appropriate, and on the Right-hand Side are Complex, Heterogeneous Systems, for Which Only Semi-Empirical Models May Be Appropriate. Ideas can readily be exchanged between the two systems, but constants derived in the simple systems can usually not be readily applied in more heterogeneous systems, and the temptation to do so must be critically examined.***

In one approach (Figure 1, left-hand side), one does not work with the heterogeneous material itself, but represents it as a combination of *reference materials* (RM), which are less heterogeneous and more easily characterized. Common RM include pure oxides, specimen clay minerals, simple polyelectrolytes, octanol, etc. Generally associated with experimental studies of RM are *mechanistic* models, which are based on fundamental principles such as surface-complexation and electric-double-layer concepts. These mechanistic models are more likely to be appropriate for these relatively homogeneous RM, for which *detailed, molecular-level structural information* can be obtained. The advantages of this experimental approach are that laboratory work is easier to control and that results can be interpreted more readily in terms of fundamental principles. Thus, the results can in principle be extended to interpret other processes such as reactivity, colloid stability, etc. The disadvantage of this approach is that the results of the experiments with RMs might in fact be irrelevant for the environmental materials that they are supposed to represent -- that is, the environmental material cannot be represented as a linear combination of its constituents (because of occlusion of one surface coating by another, for example). Furthermore, truly mechanistic models may be prohibitively complex for routine use [6].

In another approach (Figure 1, right-hand side), one can work with the *heterogeneous environmental complexing agent* (HECA) itself and use a *semi-empirical* model to represent the data. This sort of model is based on some distribution of complexation properties, since *no detailed molecular-level structural information* is available, and the processes cannot be represented in terms of fundamental interactions of these complex materials. For example, models of electrostatic interactions based on simple planar, spherical, or cylindrical geometries may be inadequate to represent the surfaces of soil particles or humic acids, just as simple 1- or 2-site models of chemical interactions may be inadequate to represent the chemical heterogeneity of these materials [7]. The advantages of this approach are that the materials are definitely relevant and that there is no necessity for an uncertain extrapolation from reference materials to the materials of interest. Potential disadvantages of this approach are that the experiments with these materials might be confounded with artifacts that are difficult to assess (such as dissolution of soluble minerals as a function of pH), and that the results may be difficult to interpret at all in terms of basic principles.

Ideally, given infinite resources, one would iterate between the left side and the right side of Figure 1. Study of the RM and the mechanistic models gives insight about possible ways to explain the behavior of the HECA, and study of the HECA generates questions to be examined by mechanistic studies with the RM. Thus, as indicated in Figure 1, iterative exchange of *ideas* between the two approaches could be quite useful. In contrast, as illustrated by the broken arrow at the bottom of Figure 1, equilibrium constants that were determined from the study of RM may not be immediately transferable to HECA, despite the fact that this transfer might appear to be an efficient way to obtain badly needed data. Evidence of this situation is the lack of direct, *ab initio* applications of metal speciation models to field systems that was mentioned in the introduction. Thus, as indicated in Figure 1, we conclude that *ideas* and *concepts* may be exchanged readily between the two approaches, but quantitative application of mechanistic models to HECA systems is still beyond our present capabilities, primarily due to the lack of detailed structural information for HECA.

2.3 The Course of a Hypothetical Study

The ideal multi-step iteration procedure outlined above may in practice be reduced to the rather abbreviated procedure outlined in Table 1. The procedure in Table 1 is based on the fundamental hypothesis that it is possible to represent the HECA as a combination or weighted sum of the RM, and may consist of the following steps. (1) First one determines the composition of the HECA, in order to establish the proper set of RM to represent the HECA. Emphasis should be placed on analytical techniques that probe the *surface* of the sorbent (rather than the bulk), since the surface characteristics are those that are relevant for complexation or adsorption properties. (2) Then one determines, in a detailed study, the binding of the target ion to the relevant RM for a wide variety of solution conditions. (3) Next, one represents the HECA as a weighted sum of the RM, and calculates the distribution of the target ion on the HECA. (4) Finally, for confirmation, one should measure the distribution of the target ion on the HECA. (5) If one would compare the initial calculated distribution with the initial measured distribution, one would probably find the agreement to be very poor. (This result would probably not appear in the literature.) To improve the agreement, one would go back to Step 1 to perform more chemical analysis of the HECA and adjust the weighting factors, and to Step 2 to perform additional experiments with the RM under more realistic conditions (e.g., add competing ions that originate from dissolution of the HECA).

Table 1. Approach to model development.

Basic Assumption

It is possible to calculate $K_{HECA} = \sum w_i K_{MX_i}$.

Procedure

1. Determine composition of HECA to establish w_i;
2. Determine K_{MX_i} (pH, salt, ...) experimentally;
3. Calculate $K_{HECA}{}^{calc} = \sum w_i K_{MX_i}$ (HECA, pH, salt, ...);
4. Determine $K_{HECA}{}^{exp}$(pH, salt, ...) experimentally;
5. Compare K^{calc} to K^{exp} and reconcile differences:
   (i) further analysis of HECA;
   (ii) further experiments with RM.

Evaluation

- Yields excellent understanding of mechanism;
- Lots of work (impractical).

Iterating through these steps may indeed lead to an excellent understanding of the system, but the amount of work generally makes it impractical. Most published studies stop at Step 3, and very, very few (we are not aware of any) go all of the way through Step 5. Furthermore, the differences between experimental and calculated binding isotherms might be reconciled by *ad hoc* addition of parameters to the model (e.g., additional site types, additional stoichiometries, etc.) so the pure "mechanistic" basis of the model is lost. Finally, there is the realization that many complex environmental sorbents may indeed be so complicated that they just cannot be represented as a combination of a small number of simple reactions; indeed, this conclusion is supported by the paucity of examples in the literature in which this approach has been brought all of the way to a truly mechanistic model that is applicable to field situations.

In view of the questions raised above, we may reassess the hypothesis that the HECA can be represented as a summation of RM, the properties of which can be described by fundamental chemical and physical processes. We now reexamine the goals for our metal speciation model and examine an alternative modeling approach.

2.4 Criteria for Model

The criteria that we use in selecting a model for heterogeneous environmental sorbents are presented in Table 2 and explained below:

Table 2. Criteria in selection of model.	
1.	Accurate interpolation over all solution conditions and thermodynamic consistency;
2.	"Small, orderly, and stable" set of adjustable parameters;
3.	Compatibility with solution speciation models;
4.	Insight into physical reaction mechanisms.

1. The foremost among the criteria are *accurate interpolation* over all solution conditions and *thermodynamic consistency* [8,9]. Here we emphasize data sets that we describe as *multidimensional*, that is, they include variations in several solution chemistry characteristics, such as p*H*, strong electrolyte concentration, total metal concentrations, etc.

   While interpolation is almost trivial in a one-dimensional system, it is more difficult in multidimensional systems, with the degree of difficulty increasing with the dimensionality of the data set. Thus, while at first glance our goal of "accurate interpolation" might seem overly cautious and unambitious, on second thought it is seen still to present a significant challenge; indeed, finding *any* model that is able to represent a multidimensional data set for a heterogeneous material over a wide range of concentrations is extremely difficult. Furthermore, this goal is consistent with our statement about the futility of trying to develop a *truly mechanistic* model for *heterogeneous environmental* complexing agents.

   Thermodynamic consistency [8,9] in a model for adsorption of two or more components is embodied in the requirement that:

$$\frac{\partial q_i}{\partial \ln a_j} = \frac{\partial q_j}{\partial \ln a_i} \qquad (2)$$

where q_i represents the total amount of component i sorbed and a_i represents the activity of component i in solution; details are discussed in the references cited. Isotherms derived from any meaningful equilibrium model must satisfy this condition.

2. We require that we obtain a "small, orderly, and stable" set of adjustable parameters. The set should be *small* to facilitate portability of the model. The set should be *orderly* to facilitate comparison of different environmental materials, that is, to provide a "fingerprint" of the binding properties. The set of parameters should also be *stable*, that is, the set of parameters should not be too sensitive to changes in the range of the experimental data or the error in the data. As will be shown, these requirements are difficult to satisfy simultaneously, and compromises have to be made.
3. We require that the model be compatible with solution speciation models. Association of metals with complex organics and surfaces, which is difficult to model, is always coupled with speciation in the simple inorganic matrix, which is comparatively easy to model.
4. We prefer models that give insight into physical reaction mechanisms, although we don't demand too much from the model in this respect due to the difficulties discussed earlier. Thus, we attempt to incorporate chemical and structural information wherever possible, and we base the model on fundamental laws such as mass action and material balance.

Finally, we note that minimizing the number of adjustable parameters -- just for the sake of minimizing the number of adjustable parameters -- is not among our criteria; we simply require a "small, orderly, and stable" set of adjustable parameters.

2.5 An Example - Approaches to Representing Humic Substances

It is generally recognized that heterogeneous environmental complexing agents (e.g., humic substances, surfaces of suspended particles) must consist of many different functional groups with a variety of different binding properties. Although these materials are in reality very complex, they have been represented in many models through various simplifications, as discussed below for the example of humic substances.

One of the first approaches to modeling the metal binding properties of complex organic matter was to base the model on real ligands of similar structure or similar equilibrium constants (e.g., salicylate, benzoate, etc.) [10,11]. While this approach appeals directly to a chemist's intuition, it is obviously somewhat artificial, and alternative approaches, which are based on hypothetical ligands with systematically distributed values of log *K*, impart more flexibility into the model without a great sacrifice in conceptual appeal (e.g., [12,13]). Hence, we consider discrete and continuous distributions of log *K*. In all of these approaches, the electrostatic energy is not expressed explicitly, but is incorporated implicitly in the values of log *K*. This approach is somewhat similar to that used routinely for simple polyprotic acids -- for example, one characterizes the simple, symmetric, diprotic oxalic acid in terms of pK_{a1} and pK_{a2}, not by a single "intrinsic" pK_a and an electrostatic correction term for the second deprotonation step.

Alternatively, models have been developed with explicit electrostatic energy terms. These models include electrostatic terms formulated with planar, cylindrical, or spherical symmetry in combination with elementary chemical reactions, for example, the complexation electric-double-layer models (e.g., [14,15]).

Out of this wide variety of possibilities, we prefer the approach based on a hypothetical ligand with a discrete-log-*K* affinity spectrum. This approach is certainly consistent with our criteria for a "small and orderly" set of parameters, compatible with solution speciation models, and yields some mechanistic insight. Furthermore, our experience has shown that this approach can be successfully extended to multidimensional data sets.

3. DEVELOPMENT OF AFFINITY SPECTRUM MODELS

3.1 Generic Semi-empirical Langmuir-based Model

The discrete log *K* spectrum approach is based on the fundamental equations of mass action and mass balance, which, for the one-dimensional case with 1:1 stoichiometry, are:

$$[\mathrm{M}]\,[\mathrm{X}_i]\,K_{\mathrm{MX}_i} = [\mathrm{MX}_i] \quad (3)$$

$$T_{\mathrm{X}_i} = [\mathrm{X}_i] + [\mathrm{MX}_i] \quad (4)$$

where [M] represents the concentration of metal, $[X_i]$ represents the concentration of free binding sites of type i, K_{MX_i} is the stability constant, $[MX_i]$ is the concentration of the bound metal, and T_{X_i} is the total analytical concentration of binding sites of type i. Activities may be used in place of concentrations in the mass action equations. This model is equivalent to a Langmuir model for each site i. In Equations 3 and 4, only metal binding is indicated, but the actual model that we use includes binding of protons and many different metals, and hence could be referred to as a competitive multisite Langmuir model. Thus, in a sense, this model has "mechanistic" features and is in principle suitable for representing competition, saturation, and comparison of energies. Furthermore, its generalization to several components is thermodynamically consistent [8,9] and it is immediately compatible with general speciation models.

3.2 Determination of the Affinity Spectrum from Experimental Data

The next point to discuss is how one actually constructs a discrete log K model from experimental data. This point is illustrated with the hypothetical adsorption data in Figure 2, which is close to a simple one-component Freundlich isotherm with a slope of approximately 0.7. The task is to find some combination of Langmuir isotherms, which are consistent with Equations 3 and 4, that reproduce the adsorption data in Figure 2. By a set of conventions discussed below, we set log K_{MX_i} = 3, 5, 7, 9, and then determine the total concentrations T_{X_i} of each of the sites i; these values with Equations 3 and 4 then constitute the model.

The values of K_{MX_i} in the affinity spectrum are determined from a loose set of rules developed through experience. We set values of log K_{MX_i} (the *spectral* window) to span the range of -log [M] in the experimental data set (the *data* window), and we set the value of Δ log K_{MX_i} = 1 or 2. The basis for setting the spectral window to span the data window can be gathered from inspection of Equations 3 and 4. Note that at log K_{MX_i} = -log [M], the value of $[X_i] = [MX_i]$; at log K_{MX_i} >> -log [M], X_i is completely saturated with M, and at log K_{MX_i} << -log [M], X_i is completely free. Thus, values of log $K_{MX_i} \approx$ -log [M] are in the range that are useful for constructing the composite isotherm; accordingly, we set the bounds of the spectral window to span the data window, which we define by:

$$-\log [M]_{max} - \gamma < \log K_{MX_i} < -\log [M]_{min} + \gamma \qquad (5)$$

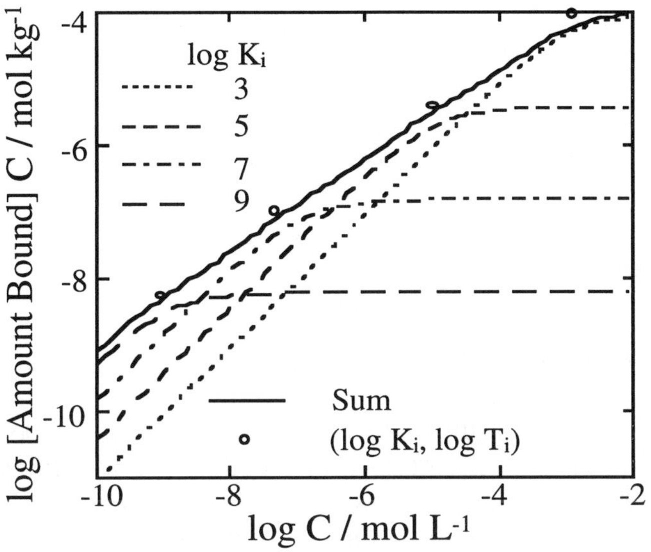

Figure 2. ***Representation of an Approximately Freundlich Isotherm (with log-log slope of 0.7) as a Sum of Langmuir Isotherms (solid line). The individual Langmuir isotherms (dotted lines) were generated from Equations 3 and 4 with log*** $K_{MX_i} = 3, 5, 7, 9$ ***and corresponding values of*** $T_i = 10^{-4.0}, 10^{-5.4}, 10^{-6.8}, 10^{-8.2}$.

where $[M]_{max}$ and $[M]_{min}$ are the maximum and minimum concentrations of M in the experimental data set, and γ is a "margin" term, the value of which is usually 1 or 2 [16,17]. If values of log K_{MX_i} are much smaller than the bounds of the data window, then $[X_i] \approx T_i$, and the isotherm is linear with a slope given by the product $K_{MX_i} T_i$. Conversely, if values of log K_{MX_i} are much larger than the bounds of the data window, the resulting isotherm is a simple constant (since $[MX_i] \approx T_i$ and is effectively constant) and the exact value of this K_{MX_i} is immaterial.

A similar simple approach can be taken for the choice of Δ log K: if Δ log K is very small, obviously a very good fit to almost any curve can be achieved, but the set of parameters is no longer "small and orderly"; in contrast, a large value of Δ log K leads to a poorer fit but a smaller and more orderly set of parameters. One can determine bounds on errors for a particular combination of Δ log K and a particular experimental isotherm, but in general, values of Δ log K = 1 or 2 have been found to be successful [16,17].

Next we discuss several examples of this approach, including models for the binding of metals to both humic substances and soil particle surfaces as a function of solution pH, salt concentration, total metal concentration, and competing cation concentration. In all of the examples, remarkable predictive properties of these models are seen. These examples involve the application of both FITEQL 3.1 [18], which is a conventional least squares optimization procedure that includes solution speciation, and a contemporary *regularized least squares* optimization technique, the use of which has been discussed recently in relation to chemical equilibrium models [16,17,19].

4. APPLICATIONS

4.1 Example Based on FITEQL

The first example illustrates the use of FITEQL to determine the binding of Co(II) to a Leonardite humic acid (LHA) as a function of salt ($NaClO_4$) concentration, pH, and total Co(II) concentration [20]. The data include an acid-base titration curve in the absence of Co(II) as shown in Figure 3a, the bound-free distribution ratio of Co(II) as a function of pH (Figure 3b), and the distribution of Co(II) at constant pH and varying total Co(II) concentration (Figure 3c), which is somewhat similar to a binding isotherm at a constant pH.

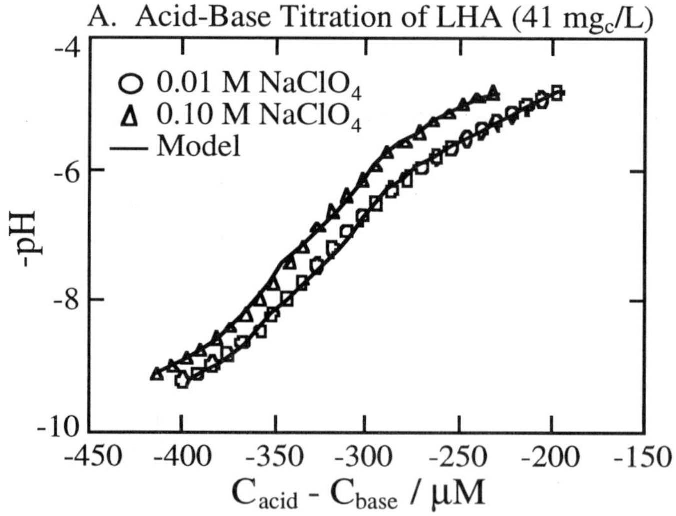

Figure 3. ***Association of Co (II) with Humic Acid (LHA) as a Function of pH, [NaClO$_4$], and Total Co(II) [20]. A. Acid-base titration of LHA.***

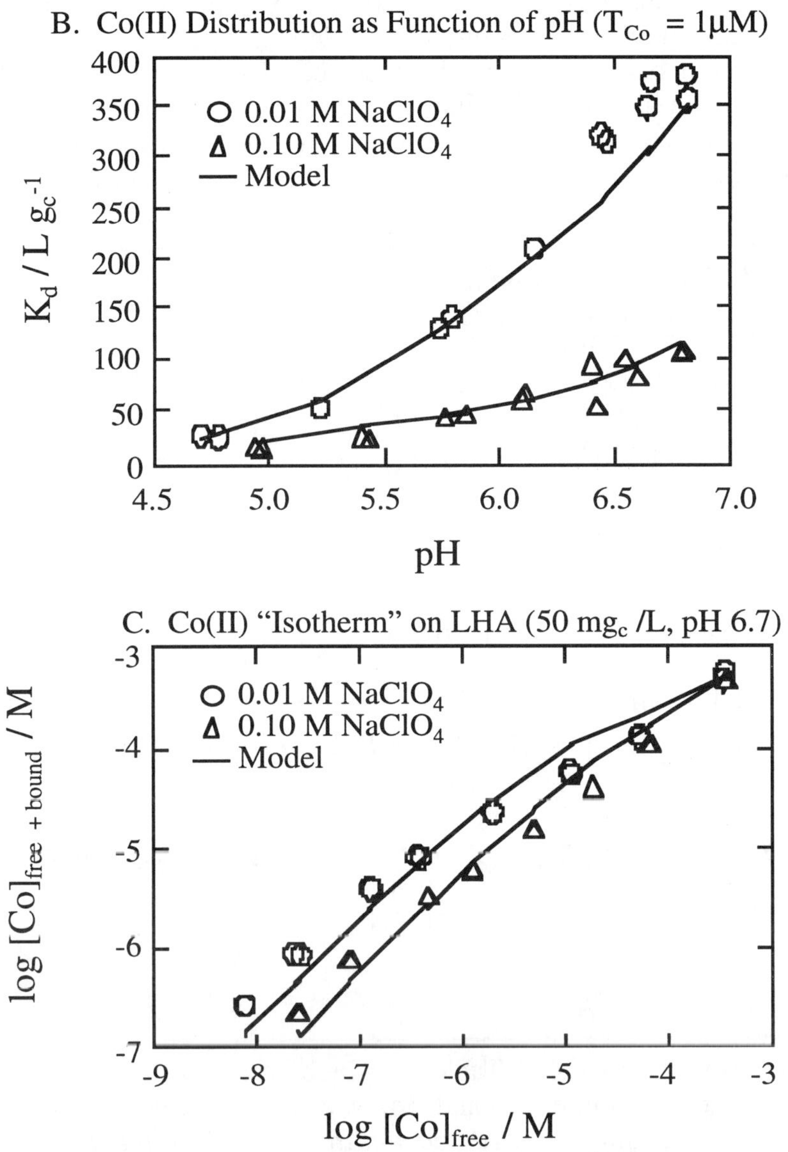

Figure 3. ***(Continued); B. Distribution ratio of Co(II) as a function of [NaClO$_4$] and pH, for total Co(II) = 1 μM, as determined by dialysis experiments; C. Distribution of Co(II) as determined by dialysis experiments as a function of [NaClO$_4$] and total Co(II), at approximately pH 6.7.***

The model was developed in two steps. First, the acidity constants and sodium binding constants were determined from the data in Figure 3a. As shown in Table 3, The log K_{HX} spectrum was fixed at values of log K_{HX_i} = 4, 6, 8, 10, to span the pH range of the data, and the acidity constants and the sodium binding constants, which incorporate the salt-concentration dependence of protonation-deprotonation behavior, were determined. The results are shown in Table 3. Then the Co(II) binding constants for these hypothetical acids were determined from the data in Figure 3b; the results are also shown in Table 3. Additional details about the model are reported elsewhere [20].

Table 3. FITEQL model for binding of Co(II) by Leonardite humic acid [20].

Reactions

$H^+ + X_i^- = HX_i$ $\quad K_{HX_i}$

$Na^+ + X_i^- = NaX_i$ $\quad K_{NaX_i}$

$Co^{2+} + X_i^- = CoX_i^+$ $\quad K_{CoX_i}$

Material balance

$$T_{X_i} = [X_i^-] + [HX_i] + [NaX_i] + [CoX_i^+]$$

Constants

Site Type	$\log K_{HX_i}$	T_{HX_i} μM	$\log K_{NaX_i}$	$\log K_{CoX_i}$
I	4	190	1.7	--
II	6	88	1.7	5.4
III	8	58	1.7	6.4
IV	10	79	1.7	--

Finally, the model in Table 3, which was developed exclusively from the data in Figures 3a and 3b, was used to calculate the Co(II) distribution in Figure 3c, *with no additional adjustable parameters*. As seen in the figure, the agreement between the model and the data is remarkably good, particularly considering that the calibration data set was based on only one total concentration of Co(II) (≈ 1-3 μM), and the range of total concentration of Co(II) in Figure 3c is 200 nM to 0.5 mM. Thus, the semi-empirical model based on a combination of Langmuir isotherms incorporates significant *extrapolative powers* in this case.

Thus, the advantages of the semi-empirical approach can be seen -- ability to represent the data and significant predictive powers (for at least this case). However, the mathematical framework of the current FITEQL Version 3.1, which functions quite satisfactorily when the ligands are well characterized and the stoichiometry is well known, is not really designed for the *ill-posed* problems associated with affinity distributions. Thus, for example, the optimization procedure diverged initially when the attempt was made to determine Co(II) binding constants for all four of the acid groups simultaneously. Of course, this difficulty was overcome by manually inserting and deleting species until convergence was achieved. In this case, with a small number of species, this tedious procedure can still be successful; however, in another case, it might not be. A better solution would be an alternative numerical procedure that effectively handles this problem automatically. Towards this end, we consider *regularized least squares* optimization with a non-negativity constraint, as discussed next.

4.2 More General Log *K* Spectra Through Regularized Least Squares

The problem of determining the affinity distribution is *ill-posed* -- that is, with the usual level of experimental noise (0.1% or more), there are in principle a large number of distributions with which the experimental data can be represented. To alleviate the numerical problems associated with this ill-posedness, *regularized* least squares techniques can be used. In these procedures, one makes an *a priori* hypothesis about how the log *K* distribution should appear -- for example, smooth, or small number of discrete sites, etc. -- and then introduces a penalty function into the weighted sum of squares that favors distributions that exhibit these properties; this procedure is outlined in Table 4, in which two regularizers are given as examples. (In the classical FITEQL solution to the Co(II)-humic acid problem cited above, a degree of "regularization" was introduced implicitly in the way the problem was set up: small number of sites with integral values of log *K*.)

The application of contemporary numerical techniques [21] for regularization has been discussed recently by Černík *et al.* [16], of which an illustration is given in Figure 4. The data set is a simulated acid-base titration curve, which was created by Gamble and Langford [22] and discussed again by Nederlof *et al.* [23]. The data were generated by simulating the titration of a mixture of four monoprotic acids, then imposing a 2% relative standard deviation on both the bound amount (q) and the free concentrations (c). The true affinity distribution (used to generate the data) and the simulated isotherm (with error imposed) are shown in Figure 4a. In Figures 4b-d, affinity spectra and residuals resulting from three different levels of regularization are shown; a value of $\Delta \log K = 0.1$ on a fixed grid was used in all cases.

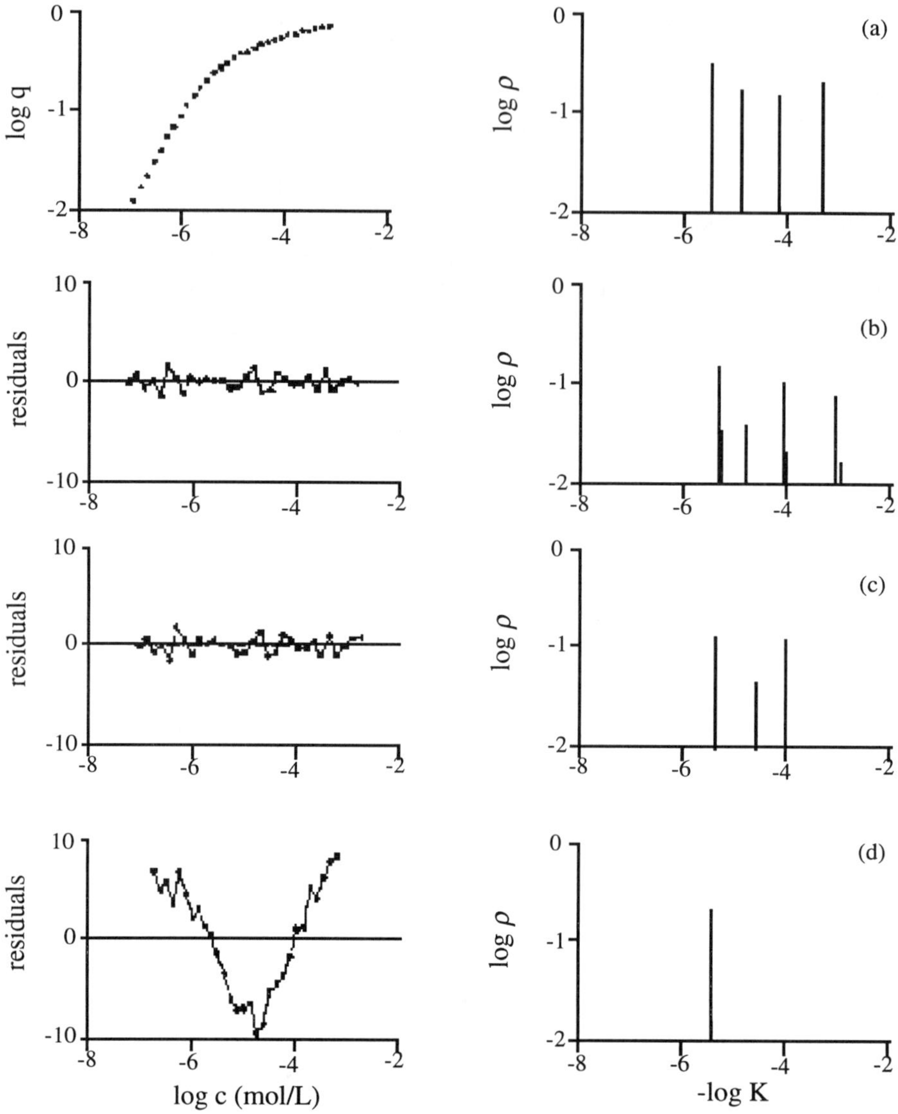

Figure 4. ***Model of Titration Curve of Four Monoprotic Acids [22] with Varying Degrees of Regularization [16]. A. The generated data set (q, mol/kg adsorbed; c mol/L in solution) with 2% RSD random error imposed on q and c, and the normalized affinity distribution with which the data set was generated; B. residuals*** $[q_i - q(c_i)] / \sigma_i$ ***and affinity distribution with no regularization; C. residuals and affinity distribution with "optimal" regularization; D. residuals and affinity distribution with "over-"regularization. Fit on a fixed logarithmic grid with*** $\Delta \log K = 0.1$.

Table 4. Regularized least squares.

- Minimize $\chi^2 + \lambda R$

$$\chi^2 = \sum (H_{calc} - H_{exp})^2 / \sigma_H^2$$

λ = weighting parameter

R = regularizing function

- Smoothness

$$R = \sum (T_{i+1} + T_{i-1} - 2T_i)^2, \qquad i = 1,...N$$

- Small number of sites

$$R = \sum T_i T_j, \qquad i > j$$

In Figure 4b the affinity distribution calculated with no regularization and only the non-negativity constraint is shown. A total of seven peaks are seen in the distribution, with four major peaks similar to those of the original spectrum, and three satellite peaks, originating from the experimental error. The residuals are relatively randomly distributed and the value of χ^2/N is 0.74, suggesting that some of the experimental error has actually been removed through the fitting procedure. In Figure 4c, the affinity distribution with "optimal" regularization is shown. The number of sites has been reduced from seven to three, and the value of χ^2/N climbs to 1.1; the residuals still appear to be randomly distributed, and the fit of the data is fully satisfactory. If the regularization parameter is further increased, the number of sites can be reduced to one, but now the value of χ^2/N increases to 26 and the residuals show a strong trend; clearly, one site is insufficient to represent the data.

Further details on the regularized least squares optimization are discussed in detail by Černík *et al.* [16]. Conclusions of that paper are: (i) distributions can be successfully regularized for smoothness or small number of sites; (ii) sometimes the "true" distribution can be recovered from generated data, but only if the error imposed on the data is unrealistically small (e.g., 10^{-9} relative standard deviation); (iii) in general, it is impossible to recover the true distribution for HECA; (iv) regularization does stabilize the inversion procedure, allowing virtually identical spectra to be derived from different realizations of the same generated data set (i.e., the same basic set with several different versions of random experimental error imposed). Two questions that remain to be addressed with respect to the application of regularized least squares are: (i) can this very generalized method be used *to represent multidimensional experimental data over a very broad range of solution conditions*? and (ii) does the semi-empirical model determined from regularized least squares *exhibit any predictive (extrapolative) properties* similar to those exhibited in the FITEQL example described above?

4.3 Application of Regularized Least Squares

The answers to these two questions can be gathered from an examination of a two-dimensional data set of Huang and Stumm [24] for the adsorption of Ca^{2+} on Al_2O_3 in 0.1 M NaCl, which is shown in Figure 5. In Figure 5a is shown the protonation-deprotonation of the surface as a function of pH, for several different total Ca^{2+} concentrations. On the basis of these data and the reaction:

$$H^+ + X_i = HX_i \tag{6}$$

(charge on X_i omitted for simplicity) and the nonelectroneutral exchange reaction:

$$H^+ + CaY_i = Ca^{+2} + HY_i \tag{7}$$

the model represented by the solid lines in Figure 5a was developed with regularization for small number of sites with integral values of log K and Δ log $K = 1$. As can be seen, the model fits the experimental data very well. The affinity spectra for the two reactions are shown in Figure 5c. At first glance, the number of adjustable parameters in the model (there are 20 of them) may seem overwhelming, but the set is "orderly" (i.e., it could be used as a "fingerprint" of properties for comparison with other materials) and the experimental data are complex -- we have found it difficult, if not impossible, to achieve such a satisfactory representation of the data with a 1- or 2-site electrostatic model. The sites reacting according to Reaction 6 are those that can be deprotonated in the absence of Ca^{2+}, while those that react according to Reaction 7 contain protons that can only be displaced by Ca^{2+}. The incorporation of all of these sites in general speciation models is handled easily in the same way that continuous Gaussian distributions have been incorporated, namely, through a separate routine written explicitly for these special sites.

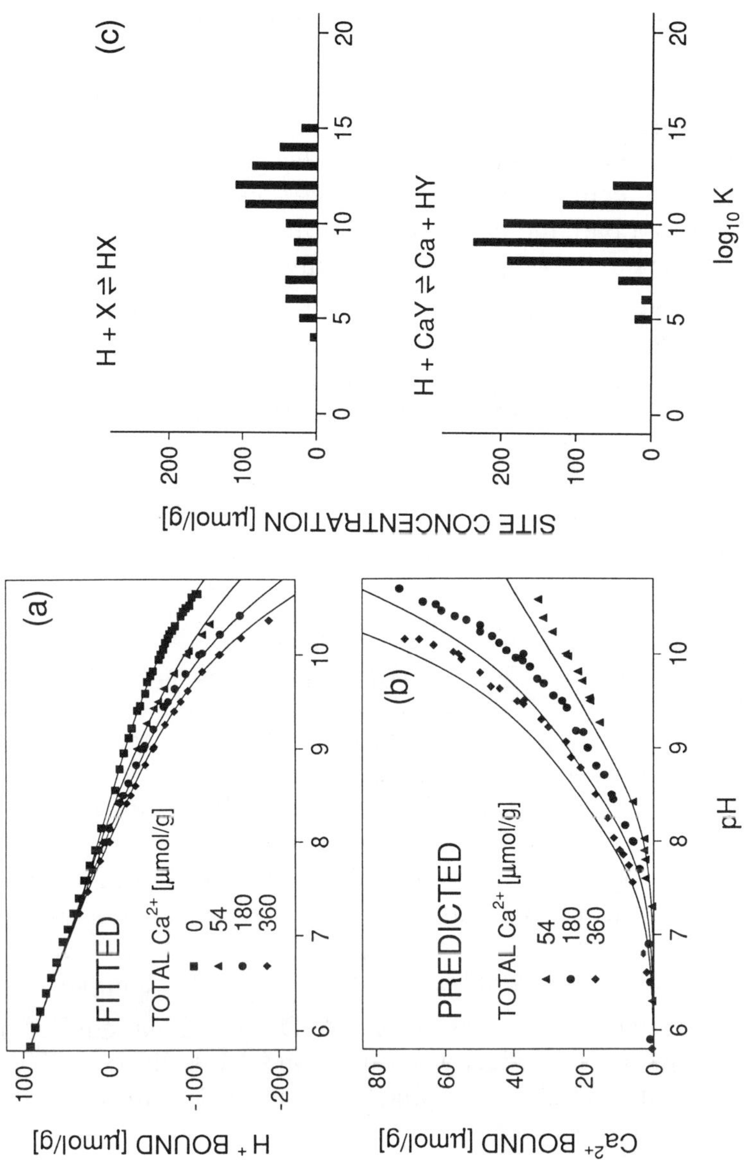

Figure 5. ***Adsorption of H^+ and Ca^{2+} to Al_2O_3 in 0.1 M NaCl as a Function of Ca^{2+} and H^+ in Solution. Experimental data for an alumina concentration of 2.8 g/L measured by Huang and Stumm [24]. A. H^+ adsorption data (points) as a function of pH for different total Ca^{2+} concentrations and the best fit with the model(line). B. Ca^{2+} adsorption data as a function of pH for different total Ca^{2+} concentrations and predictions of the model (line). C. Affinity distributions for the reactions in the model.***

Next the predictive powers of this semi-empirical model are tested. Experimental data for Ca^{2+} binding as a function of pH are presented in Figure 5b, along with the prediction calculated from the model that was calibrated solely on the data in Figure 5a -- that is, the only basis for the model of Ca^{2+} binding is the proton displacement in Figure 5a. As can be seen, the agreement between the model and the experimental data in Figure 5b is not bad, indicating that the semi-empirical model is at least mechanistic enough to capture intrinsically the $Ca^{2+} - H^+$ competition.

Thus, our answer to the two questions posed above are yes, the semi-empirical models can be used to represent complex data sets, and yes, they do have extrapolative capabilities. Although only one example was discussed above, many other examples of both properties are discussed in related papers.

4.4 Compromises Between Current Versions of FITEQL and the Contemporary Approach

Finally we can compare the compromises between the classical FITEQL approach and the new regularized least squares approach, as they are both currently implemented. For this comparison we use two datasets for the adsorption of a cationic surfactant (dodecylpyridinium, DP) on two subsoil materials as a function of salt concentration, as shown in Figures 6 and 7. This example illustrates the use of the modeling approach with 1-dimensional data over an exceptionally wide concentration range, with DP as a "model" metal ion, exhibiting no complexation reactions in solution. The original FITEQL interpretation was published several years ago by Brownawell *et al.* [25], and we present here the regularized least squares version for comparison.

First consider the adsorption of DP on Lula aquifer material (Figure 6), which is basically a silica sand naturally coated with iron oxide [25]. The FITEQL interpretation was made with three sites, two for the exchange reaction:

$$R^+ + NaX_i = RX_i + Na^+ \tag{8}$$

where R represents the organic cation, DP, and one site Y for the ion-pair adsorption reaction:

$$R^+ + Cl^- + Y_i = RClY_i \tag{9}$$

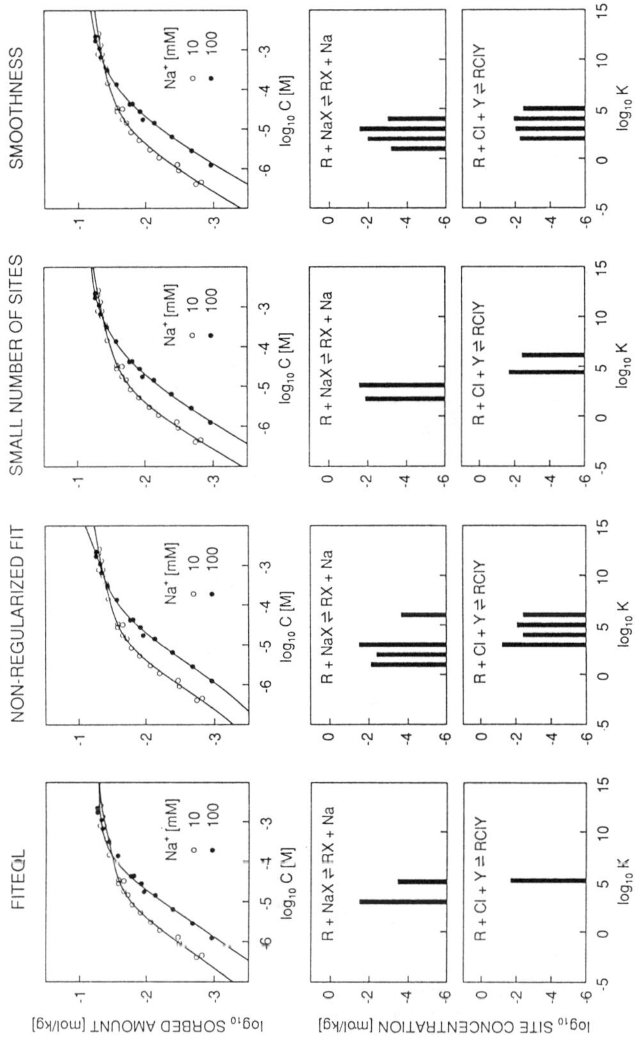

Figure 6. ***Adsorption of DP to Lula Aquifer Material by Brownawell et al. [25]. The fit of the model to the data and the corresponding affinity spectra are shown for the models determined by (A) FITEQL [18]; (B) nonregularized least-squares with a non-negativity constraint on a fixed grid with Δ log K = 1 [19]; (C) small number of sites regularization with a non-negativity constraint and a minimum of interval between successive sites of Δ log K = 1 [19]; and (D) smoothness regularization with a non-negativity constraint on a fixed logarithmic grid with Δ log K = 2 [19].***

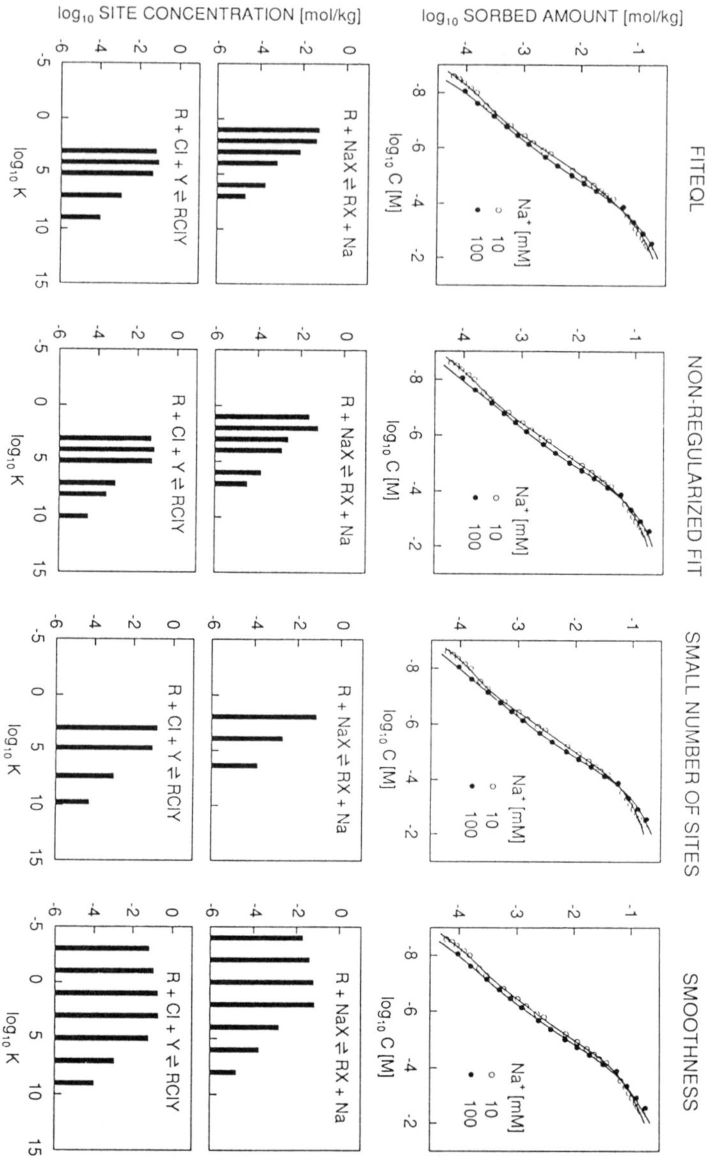

Figure 7. Adsorption of DP to Sediment Material EPA-12 by Brownawell et al. [25]. The fit of the model to the data and the corresponding affinity spectra are shown for the models determined by (A) FITEQL [18]; (B) nonregularized least-squares with a non-negativity constraint on a fixed grid with $\Delta \log K = 1$ [19]; (C) small number of sites regularization with a non-negativity constraint and a minimum of interval between successive sites of $\Delta \log K = 1$ [19]; and (D) smoothness regularization with a non-negativity constraint on a fixed logarithmic grid with $\Delta \log K = 2$ [19].

The FITEQL affinity spectrum (Figure 6) was determined by initially establishing a fixed log K spectrum that spanned the experimental data domain, and then systematically deleting and inserting components until convergence was achieved. The procedure is time consuming, but the result is satisfactory: a quite acceptable fit based on three sites, for which the total concentrations are adjustable parameters.

For comparison, three interpretations of the data with the regularized least squares techniques were made (Figure 6). If a nonregularized fit is used, the fit is better (quite obvious on a linear scale but not so obvious on the logarithmic scale), but now a total of eight sites are required. Furthermore, there seems to be some curvature at the extrema that are not obvious in the FITEQL model. If the solution is regularized for a small number of sites, the fit is satisfactory and only four sites are required; also the curvature at the extrema disappear. If one regularizes for smoothness, again more sites are required but the fit is excellent and the curvature at the extrema also disappears.

A similar result is observed for the various models and the data for adsorption of DP on another sediment, EPA-12, as shown in Figure 7. In this case, the data span almost seven orders of magnitude, and the effect of salt is not so great. The FITEQL fit requires eleven sites and is quite satisfactory. One might question the need for eleven sites, but the domain of the experimental data is very large. The log K spectra obtained for the nonregularized fit is remarkably close to the FITEQL spectrum, mutually verifying both methods. However, the fit regularized for the small number of sites is virtually identical to the nonregularized version, but only seven sites are required. Finally, the fit regularized for smoothness requires fourteen sites. Again, on a linear scale, the fit regularized for smoothness generates the smoothest isotherm, which may be closest to the real isotherm.

This brief excursion through a fitting exercise supports the following general conclusions: (i) the FITEQL method leads to results similar to those of the non-regularized least squares method, where the FITEQL method is based on trial-and-error substitution of species and the non-regularized least squares method is fully automatic; (ii) the regularization for a small number of sites leads to approximately the same or a smaller number of sites; (iii) the optimization for smoothness leads to the most aesthetically and intuitively appealing isotherms, although with a greater number of sites (and adjustable parameters).

Further to be recognized is the fact that the DP data (Figures 6 and 7) are one-dimensional; the Ca^{2+} / H^+ data are two-dimensional, and the Co(II)/ humic acid/ H^+ data are three-dimensional. At the current time, FITEQL can *in principle* handle 1-, 2-, 3-, n-dimensional data, and seems to be limited only by the user's patience to solve manually ill-posed problems by inserting and deleting components; *in practice* FITEQL appears to be limited to 2- or 3-dimensional problems due to the aforementioned limits on the user's patience. A program to solve 1-dimensional problems by regularized least squares (QUASI) is available [19]. Two-dimensional problems can be solved routinely in our laboratories, and methods for three- and n-dimensional problems are still under development.

Acknowledgments: This research was supported by the Subsurface Science Program, U. S. Department of Energy under contract DE-AC06-76RLO 1830 (Co-Contaminant Chemistry Subprogram) and the Swiss National Science Foundation.

REFERENCES

1. Waite, T.D., in *Trace Element Speciation: Analytical Methods and Problems*, G.E. Batley, Ed., pp. 117-184. CRC Press (Boca Raton, FL 1989).

2. Bassett, R.L. and D.C. Melchior, in *Chemical Modeling of Aqueous Systems II*, ACS Symposium Series 416, D.C. Melchior and R.L. Bassett, Eds., pp. 1-14. American Chemical Society (Washington, DC 1990).

3. Smith, K.S., "Factors Influencing Metal Sorption onto Iron-Rich Sediments in Acid-Mine Drainage," Ph.D. Thesis, Colorado School of Mines (Golden, CO 1991).

4. Loux, N.T., D.S. Brown, C.R. Chafin, J.D. Allison, and S.M. Hassan, "Chemical Speciation and Competitive Cationic Partitioning on a Sandy Aquifer Material," *Chem. Spec. Bioavail.* 1:111-125 (1989).

5. Müller, B. and L. Sigg, "Interaction of Trace Metals with Natural Particle Surfaces: Comparison Between Adsorption Experiments and Field Measurements," *Aquatic Sci.* 52:75-92 (1990).

6. Bashford, D. and M. Karplus, "pKa's of Ionizable Groups in Proteins: Atomic Detail from a Continuum Electrostatic Model," *Biochem.* 29:10219-10225 (1990).

7. Westall, J.C., "Modeling of the Association of Metal Ions with Heterogeneous Environmental Sorbents," *Materials Research Soc. Symp. Proc.* 353:937-950 (1995).

8. Franses, E.I., F.A. Siddiqui, D.J. Ahn, C-H. Chang and N-H.L. Wang, "Thermodynamically Consistent Equilibrium Adsorption Isotherms for Mixtures of Different - Size Molecules," *Langmuir* 11:3177-3183 (1995).

9. Kembal, C., E.K. Rideal and E.A. Guggenheim, "Thermodynamics of Monolayers," *Trans. Faraday Soc.* 44:948-954 (1948).

10. Stumm, W. and P. Brauner, "Chemical Speciation," in *Chemical Oceanography*, 2nd ed., J.P. Riley and G. Skirrow, Eds., pp. 173-279. Academic Press (New York, NY 1975).

11. Morel, F.M.M., J.C. Westall, C.R. O'Melia and J.J. Morgan, "Fate of Trace Metals in Los Angeles County Wastewater," *Environ. Sci. Technol.* 9:756-761 (1975).

12. Eberle, S.H. and W. Feuerstein, "On the pK Spectrum of Humic Acid from Natural Waters," *Naturwissenschaften*, 66:572-573 (1979).

13. Leuenberger, B. and P.W. Schindler, "Application of Integral pK Spectrometry to the Titration Curve of Fulvic Acid," *Anal. Chem.* 58:1471-1474 (1986).

14. Nederlof, M.M., J.C.M. De Wit, W.H. Van Riemsdijk and L.K. Koopal, "Determination of Proton Affinity Distributions Discharge for Humic Substances," *Environ. Sci. Technol.* 26:846-856 (1993).

15. Bartschat, B.M., S.E. Cabaniss and F.M.M. Morel, "Oligoelectrolyte Model for Cation Binding by Humic Substances," *Environ. Sci. Technol.* 26:284-294 (1992).

16. Černík, M., M. Borkovec and J.C. Westall, "Regularized Least-squares Methods for the Calculation of Discrete and Continuous Affinity Distributions for Heterogeneous Sorbents," *Environ. Sci. Technol.* 29:413-425 (1995).

17. Borkovec, M., U. Rusch, M. Černík, G.J.M. Koper and J.C. Westall, "Affinity Distributions and Acid-base Properties of Homogeneous and Heterogeneous Sorbents: Exact Results versus Experimental Data Inversion," *Coll. Surfaces A.* 107:285-296 (1996).

18. Herbelin, A.L. and J.C. Westall, "FITEQL. A Computer Program for Determination of Chemical Equilibrium Constants, Version 3.1," Report 94-01, Department of Chemistry, Oregon State University (Corvallis, OR 1994).

19. Černík, M., U. Rusch, and M. Borkovec, "QUASI - Version 1.0," Institute of Terrestrial Ecology, Soil Chemistry, Swiss Federal Institute of Technology (Grabenstrasse 3, 8952 Schlieren, Switzerland 1995).

20. Westall, J.C., J.D. Jones, G.D. Turner and J.M. Zachara, "Models for Association of Metal Ions with Heterogeneous Environmental Sorbents. 1. Complexation of Co (II) by Leonardite Humic Acid as a Function of pH and $NaClO_4$ Concentration," *Environ. Sci. Technol.* 29:951-959 (1995).

21. NAG Fortran Library, Mark 15, Numerical Algorithm Group, Ltd, Wilkinson House, Jordan Hill Road, Oxford, UK OX2 8DR.

22. Gamble, D.S. and C.H. Langford, "Complexing Equilibria in Mixed Ligand Systems: Tests of Theory with Computer Simulations," *Environ. Sci. Technol.* 22:1325-1336 (1988).

23. Nederlof, M.M., W.H. Van Riemsdijk and L.K. Koopal, "Comparison of Semianalytical Methods to Analyze Complexation with Heterogeneous ligands," *Environ. Sci. Technol.* 26:763-771 (1992).

24. Huang, C.P. and W. Stumm, "Specific Adsorption of Cations on Hydrous γ-Al_2O_3," *J. Coll. Interfac. Sci.* 43:409-420 (1973).

25. Brownawell, B.J., H. Chen, J. Collier and J. Westall, "Adsorption of Organic Cations to Natural Materials," *Environ. Sci. Technol.* 24:1234-1241 (1990).

11

PARTITIONING OF METALS TO SUSPENDED PARTICLES

Laura Sigg
Swiss Federal Institute for Environmental Science and Technology (EAWAG),
CH - 8600 Dübendorf, Switzerland

1. INTRODUCTION

The partition of metals between solution and suspended particles is a subject of interest with numerous applications concerning the fate and the effects of metal ions in aquatic systems. The effects of metals on biota are related to the dissolved metal concentrations, and especially to the free metal ion concentrations [1]. The dissolved concentrations are determined by the interactions of metal ions with suspended solids by adsorption, precipitation, uptake; the concentrations of free metal ions depend on the overall speciation, which includes the interactions with suspended particles, as well as the complexation by dissolved ligands. It would be useful, with respect to the interactions with biota, to be able to predict the changes in dissolved concentrations under varying conditions, e.g., of pH and of suspended solids.

The transport behavior of metals is strongly dependent on their partition between the dissolved and the particulate phase; this partition has consequences for the fate of elements in rivers, lakes, oceans and at the boundaries between these systems. Particulate bound metals will be deposited together with the sediments, e.g., in estuaries, harbor areas etc. Contamination of sediments with metals in areas of high sediment deposition is a problem that causes some concern, for example in the River Rhine. Partition of metals between suspended solids and water is also an essential factor determining the concentrations of metals in the water column of lakes and of the oceans [2-4].

For trace metal ions, binding to the particulate phase is mostly determined by surface reactions, whereas precipitation is in most cases an unlikely process at the low natural concentration levels. Binding to the solid phase depends therefore on the surface chemical properties of the particles (specific surface area, number and type of surface groups, surface charge); the conditions in solution (pH, ligands) also influence the binding reactions on surfaces.

I will attempt in this paper to show relationships between field observations of partition and the surface chemical properties of the suspended particles. The following questions will be considered:

i) How similar is the partition of metals in various environments? A similar partition may be expected if the composition of particles, as well as the solution composition are similar.

ii) Can the partition of metals between water and suspended particles be predicted on the basis of the composition of particles of known adsorption properties and of the composition of the solution?

The surface chemical properties of various solid materials, which play an important role as components of natural suspended particles, are discussed. Special emphasis is given to the interactions of metal ions with algae.

2. SIGNIFICANCE OF PARTITION COEFFICIENTS

Partition coefficients, generally called K_D, have been a very popular way of defining the partition between water and suspended particles; they indicate the enrichment in the solid phase in comparison to the solution. K_D are collective parameters, which are influenced by a number of different chemical interactions and which implicitly include many variables.

The partition coefficient is generally defined as the ratio of the concentrations in the solid and in the dissolved phase (in a conceptual definition the truly dissolved concentration):

$K_D = c_s/c_w$, where c_s is the concentration in the solid phase (moles/kg or moles/g) and c_w is the concentration in the dissolved phase (moles/liter or moles/m^3). In this paper the units used for K_D are moles kg^{-1}/moles L^{-1} (L kg^{-1}).

Assuming that the partition coefficient is determined by the interactions of metal ions with surfaces and with ligands in solution, K_D can be represented in terms of thermodynamic constants.

If binding of a metal Me^{2+} to the solid phase occurs by surface adsorption to an oxide-type surface, a reaction of the following general type may be formulated [5]:

$$>S\text{-}OH + Me^{2+} \leftrightarrows >S\text{-}O\text{-}Me^{+} + H^{+} \qquad (1)$$

where >S-OH means a surface hydroxyl group. A reaction stoichiometry yielding 2 H^+ per adsorbed metal ion is also possible.

In solution, a metal may be present in a number of different species, due to complexation with inorganic (e.g., CO_3^{2-}) and organic ligands (represented here by L_i):

$$[Me]_{sol} = [Me^{2+}] + \Sigma\,[Me(OH)_n] + [MeCO_3] + \Sigma[MeL_i] \qquad (2)$$

According to these two types of reactions, the partition coefficient may then be formulated as [6]:

$$K_D = \frac{\{>SO-Me^+\}}{[Me^{2+}]+\Sigma[Me(OH)_n]+[MeCO_3]+\Sigma[MeL_i]}\ \frac{mol\,kg^{-1}}{mol\,L^{-1}} \qquad (3)$$

or using the corresponding stability constants:

$$K_D = \frac{K_1^s\{>SOH\}[H^+]^{-1}}{1+\Sigma\beta_H^*[H^+]^{-n}+\beta_c[CO_3^{2-}]+\Sigma\beta_i[L_i]} \qquad (4)$$

where K_1^s is a surface complexation constant for a reaction of type (1) and β_H^*, β_c and β_i are the respective stability constants for hydroxo, carbonato and organic complexes.

In the case of natural particles, the metal ions bound to the solid phase would represent a sum of adsorbed species on various surface group types, with the general stoichiometry:

$$n >S\text{-}OH + Me^{2+} \leftrightarrows (>S\text{-}O)_n\text{-}Me^{(n-2)+} + nH^{+} \qquad (5)$$

It is obvious from the formula (4) that the partition coefficient depends on pH, on the type of surfaces present and their density of surface groups, and on the ligands present in solution. In a simple case complexation in solution is counteracting binding to particles; the presence of strong ligands in solution tends thus to lower the partition coefficient.

Partition coefficients can be determined from field studies including metal measurements in the dissolved and in the solid phase. They can also be determined in laboratory experiments, e.g., by adsorption experiments with natural material under controlled conditions. In all experimental determinations of K_D, the operational definition of the dissolved phase is a crucial problem.

3. EXAMPLES OF DATA ON PARTITION OF METALS BETWEEN WATER AND SUSPENDED PARTICLES

By comparing the partition of metals in various environments, some general tendencies may be recognized. If the binding to particles is mostly determined by binding to particle surfaces of similar composition, similar partition coefficients may be expected within a wide range of concentrations. Table 1 summarizes field data from various, mostly riverine environments, as well as from some estuaries [7-17]. The rivers and estuaries represent a wide range of levels of contamination, as indicated by the concentrations of metals in suspended solids. River Glatt is an example of a strongly polluted small river, which receives large volumes of sewage treatment plant effluents. The four stations on the River Rhine (Rekingen, Village-Neuf, Coblence and Lobith [8]) represent increasing loading with sewage and industrial effluents. At the other end of the scale, the Lena estuary [17] and some of the Texas estuaries [16] are rather pristine environments. In all these cases, pH is expected to be circumneutral, in the range of about 7.5 - 8.5; pH is however not well documented in every case.

The comparison of the partition coefficients in these various cases shows that there is quite a high variability, even between samples taken at the same site, but at different times. According to the data in this table, there is no obvious relationship between concentrations in the suspended particles and partition coefficients; data at various levels of metal concentrations give similar results. Typical ranges of K_D are obtained for various elements, e.g., for Pb, Cu, Zn, which are rather consistent in the various environments, even comparing rather pristine sites with polluted ones. This similarity of K_D implies that similar interactions of metal ions with the surfaces of particles determine K_D at various sites.

Table 1. Metal concentrations in suspended solids and partition coefficients (ranges of concentrations and of log K_D are given, mean values in parentheses).

Sampling sites	Zn µg/g	log K_D (Zn) L/kg	Pb µg/g	log K_D (Pb) L/kg	Cu µg/g	log K_D (Cu) L/kg	Ref.
Glatt, Fällanden	85-800 (212)	4.8-5.5	20-100 (45)	5.0-6.7	23-190 (72)	4.1-5.1	[7]
Glatt, Rümlang	255-1380 (559)	4.3-5.0	126-442 (224)	5.5-6.3	65-320 (130)	4.4-5.0	[7]
Glatt, Rheinsfelden	480 -1120 (699)	4.0-5.0	87-280 (188)	5.6-6.3	65-200 (118)	4.5-4.8	[7]
Rhine, Rekingen	125-330 (192)	4.7-5.5	37-120 (60)	5.1-5.6	25-60 (42)	4.6-5.4	[7,8]
Rhine, Village-Neuf	220-1370 (323)	4.7-5.3	17-300 (65)	4.8-5.5	50-260 (68)	4.4-5.0	[7,8]
Rhine, Coblence	240-320	4.8	70-76	5.2	63-70	4.4	[8]
Rhine, Lobith	380 -590	5	102-139	5.4	65-100	4.6	[8]
Rhone	25-151	4.4-5.0	35-111	5.4-6.3	13-191	4.7-5.2	[9,10]
Gironde			60-70	5.5-5.8			[11]
Scheldt estuary			60-200	6.0-6.4	26-137	4.4-5.3	[12]
St. Lawrence	230-437	5.7-5.8	37-198	6.5-6.8	52-140	5.0-5.5	[13]
Mississippi		5					[14]
English estuaries		4.0-5.0		5.0-7.0		4.0-5.0	[15]
Texas estuaries	20 -200	4.0-5.4	10-50	5.0-6.0	10-60	3.5-4.8	[16]
Lena estuary	90-181	5.3-5.7	16-30	5.9-6.7	24-48	4.6-5.0	[17]

Another set of data on partition coefficients of various metals has been collected by Tessier [18]; this data set (cf. literature cited in [18]) includes field data of freshwater and marine environments, as well as laboratory data with various types of natural particles. These data demonstrate clearly the pH dependence of the partition, with increasing K_D with pH. The range encountered in these various environments is at circumneutral pH consistent with the range shown in Table 1.

Some general tendencies can be derived from the data found in the literature for K_D, according to Table 1, as well as to Tessier [18]:

i) the binding of metals to particles increases with pH, according to a pH dependence in agreement with reactions of the type (Equation 1).

ii) K_D at comparable pH decrease generally in the order: Pb > Zn > Cd, Cu.

These tendencies may be compared to the binding of metals to well-defined surfaces, such as iron oxides. Binding of cations to amorphous iron oxide can be calculated on the basis of the stability constants compiled by Dzombak and Morel [19]. Applying these stability constants results in binding of these metals in the order: Pb > Cu > Cd, Zn. Similar results are obtained with other types of surfaces such as SiO_2 and TiO_2 [20]. These results are consistent with the correlation between stability constants for the surface complexation of metals on oxides and the hydrolysis constants of metals [20]. The somewhat different order obtained under natural conditions, especially for Cu, probably indicates the influence of organic ligands in solution, which according to Equation 4, would lower the partition coefficients.

In considering these field-derived data it must, however, be recognized that all experimental determinations of partition coefficients are operational approximations of the conceptual definition given above, especially with regard to the distinction between truly dissolved and particle-bound species; many experimental difficulties are encountered. A better consideration of these analytical problems may result in a better comparability of various data sets.

Analytical problems are associated with the determination of low levels of dissolved metals. Ultraclean techniques need to be used in sampling and sample preparation. Some of the partition coefficients may be underestimated because the dissolved concentrations are too high due to contamination or high detection limits (e.g., data on the Rhine River for Pb, Table 1).

The conditions in solution (pH, ionic strength, major ion composition, DOC), which are important in order to interpret the partition coefficients, are often incompletely reported in determinations of partition coefficients. More information on, e.g., the complexation by organic ligands in solution, would help in the interpretation of the data and in the comparison of various cases.

As mentioned above, a crucial analytical problem is the clear distinction between dissolved and particulate species. The usual method of filtering through 0.45 μm filters for determining dissolved concentrations includes the colloidal particles (e.g., in the range 0.010 - 0.45 μm) in the operationally defined dissolved fraction. This method therefore tends to underestimate the partition coefficients, with respect to truly dissolved concentrations. Some recent studies explicitly take into account the colloidal fraction and show that colloidal species may represent a large fraction of the <0.45 μm metal species [16,21]. The so-called "particle concentration effect" means decreasing partition coefficients with increasing particle concentrations and is contrary to thermodynamic prediction. This effect has recently been documented for the partition of metal ions in freshwaters and in estuaries [16,22]. This effect can probably be explained by the influence of colloidal particles, which by being included in the operationally defined dissolved fraction, result in an apparent decrease of the partition coefficients [16,22-24].

4. SURFACE CHEMICAL PROPERTIES OF SUSPENDED PARTICLES

The field data mentioned above indicate the significance of binding to surfaces for the binding of metals to suspended particles. It is therefore of great interest to define the surface chemical properties of the natural assemblage of particles. Two different approaches can be taken for this purpose: i) the surface chemical properties of the natural particles can be studied in laboratory experiments, in which the solution conditions are controlled; ii) the bulk composition of the particles can be used to estimate the surface properties, by taking into account the properties of various well-defined solid phases.

4.1 Experimental Approaches to Surface Chemical Properties of Natural Particles

In order to determine the surface properties of particles, laboratory experiments can be performed, in which particles from natural environments are isolated and adsorption is studied under defined conditions in the solution. A number of experimental difficulties are, however, associated with such experiments. The particles collected by centrifugation or by filtration usually do not include the colloidal fraction, which might be important for adsorption. The natural particles represent an assemblage of components with various properties, including also various surface groups within one component. Depending on the experimental conditions used, different results may be obtained, as different functional groups are sensed. The binding affinity of metal ions is expected to vary with the ratio of concentrations in solution to concentrations of surface groups, as the strongest binding groups are first occupied. Problems arise also when natural particles are submitted to pH conditions different from the original pH of their environment. Dissolution reactions (e.g., of $CaCO_3$) may occur that change the composition and reactivity of particles.

In spite of these difficulties, some studies are found in the literature, in which adsorption parameters of natural particles have been determined (e.g., [14,25-30]. These studies illustrate generally the pH dependence of adsorption, and the differences between various metal ions with respect to their affinity towards surfaces of natural particles. The constants derived from such experiments are, however, always conditional for particular experimental conditions. The study by Müller and Sigg [30] used particles collected from a river system at different times and sites. Conditional stability constants at constant pH were derived for Zn and Pb, which showed the higher affinity of Pb over Zn for the particle surfaces. Consistent results were obtained in this case for samples collected at different times; the adsorption capacities obtained are shown in Table 3, together with other examples of solid phases. The results obtained by Shiller and Boyle [14] on particles from the Mississippi River illustrate the strong pH dependence of adsorption of metal ions.

It remains a difficult problem to relate experimental data on conditional stability constants of natural particles to their composition, and to evaluate the relative significance of the various components such as organic matter, iron and manganese oxides, silicates, and calcium carbonate.

4.2 Binding to Algae: Adsorption and Uptake

The role of phytoplankton for the regulation of trace elements has been demonstrated in the oceanic environment [1,31]. Algae also play an important role in regulating the concentrations of metal ions in the water column of lakes [2,3]. The interactions of algae with metal ions of course include not only surface interactions, but also uptake of trace elements either as essential elements (e.g., Cu, Zn), or as elements without known biological function (e.g., Cd). The surfaces of algae provide various types of surface groups, on which metal ions may be adsorbed; these complexing surface groups include carboxyl, amino groups and sulfide groups in proteins at the cell surfaces.

In recent studies in our laboratory, the interactions of protons and of Cu and Zn with the surfaces of various algae have been studied [32,33]. Some of the results are summarized in Table 2. The titration of algae surfaces (*Chlamydomonas reinhardtii*) with protons could be interpreted in terms of two types of surface groups with pK_a values of about 5 and 9; these two types of surface groups may be interpreted as carboxyl and amino groups in proteins. Adsorption experiments of Cu were performed at low levels of Cu^{2+} (10^{-12} – 10^{-10} M), by using metal ion buffers. Under these conditions, high conditional stability constants and low numbers of surface groups were obtained. The number of specific binding groups for Cu is much lower than the total number of groups obtained with protons; the Cu binding groups at these low levels are very specific sites with high affinity for Cu. A possible specific structure for Cu binding at algae surfaces could be histidine, as indicated by ESR spectra [32,34].

Table 2. Maximum adsorption capacities (Γ_{max}, in mol/g dry weight) and conditional stability constants of algae surfaces with the indicated cations.

Algae		pH	log K	Γ_{max} mol/g	Ref.
Cyclotella cryptica	Cu	6.9	11.8	7.5×10^{-7}	[32]
Chlamydomonas reinhardtii	Cu	6.9	11.3	6.2×10^{-6}	[32]
Chlamydomonas reinhardtii	Cu	4.1	8.3	3.4×10^{-6}	[32]
Chlamydomonas reinhardtii	H		5.0 / 9.0	9×10^{-4}	[32]
Scenedesmus suspicatus	Cu	7.9	11.1	2.9×10^{-6}	[33]
Scenedesmus suspicatus	Zn	7.9	6.5	1.7×10^{-5}	[33]
Vaucheria	Cu	4.5	3.2	4.6×10^{-4}	[35]

In studies using higher concentrations of Cu, higher numbers of binding groups on algae surfaces were obtained (e.g., [35]). The less specific groups are being titrated at high metal concentrations; for example carboxyl groups with lower affinities are also occupied.

Table 3. Maximum adsorption capacity of various types of surfaces, as measured by adsorption of the indicated cations. (Adsorption capacity for manganese oxide is from laboratory experiments with natural MnO_2 particles collected from Lake Zurich).

Surface	Adsorbed ion	Adsorption capacity mol/g	Ref.
FeOOH (goethite)	H^+	1.35×10^{-4}	[36]
	Pb^{2+}	6×10^{-5}	
$Fe(OH)_3$ (am.)		2×10^{-3}	[19]
MnO_2 (from Lake Zurich)	Cu^{2+}, Mn^{2+}	1.8×10^{-3}	[37]
$CaCO_3$		1.8×10^{-5}	[38]
Algae (Chlamydomonas)	H^+	9×10^{-4}	[32]
	Cu^{2+}	6×10^{-6}	
Susp. particles from River Glatt	Pb^{2+}, Zn^{2+}	$5 \times 10^{-6} - 1 \times 10^{-5}$	[36]

Uptake of metal ions has of course also to be considered in addition to adsorption, in order to evaluate the total bound metal ions. In a study showing the dependence of the cellular content of Cu as a function of $[Cu^{2+}]$, both intracellularly and extracellularly bound Cu increased with increasing $[Cu^{2+}]$ [33]. However, the extracellularly bound Cu was lower than intracellularly bound Cu over the range of $[Cu^{2+}] = 10^{-14} - 10^{-7}$, in cells which had been allowed to grow in culture media with these concentrations. A different distribution between intracellular and extracellular metal ions may be obtained in experiments in which the cells are in contact with higher concentrations of metal ions for only a short time. In the case of Zn, the uptake was regulated over the range $[Zn^{2+}] = 10^{-14} - 10^{-8}$ M; only a slight increase in the intracellular content was observed over this range and very little adsorbed Zn was present for $[Zn^{2+}] < 10^{-9}$ M.

These studies show that algae may be very important both in taking up metal ions and in binding them to surfaces. Very specific functional groups, which tightly bind metal ions, are present on surfaces. Due to the regulation of uptake, the content of metal ions in algae is not always related to the concentrations in solution in a simple manner. Therefore, the simple concept of a partition coefficient must be used with caution in systems which are dominated by biological interactions.

4.3 Composition of Natural Particles and Surface Functional Groups

Natural particles are typically composed of organic material, either in intact organisms or as degradation products of organisms, of clay minerals, of calcium carbonate, and of iron oxides and manganese oxides. All these different solid components may be involved in the binding of metal ions. They differ in the types and numbers of functional groups on their surfaces, and in the affinity of these surface groups for metal ions. Some examples of the number of surface groups on various defined solid phases and on biological material are given in Tables 2 and 3. High numbers of surface groups are observed on amorphous solid phases (amorphous $Fe(OH)_3(s)$, manganese oxides), which are characterized by high specific surface areas. Calcium carbonate is usually present with relatively low specific surface area and, in contrast, appears to bear typically lower numbers of surface groups. Organic material contains various types of functional groups with different affinity for metal ions. These differences are evident, e.g., for algae, for which a small number of specific surface groups with high affinity for Cu is present, in contrast to a higher number of protonable groups with smaller affinity towards metal ions. The values for suspended particles in the River Glatt were obtained in adsorption experiments under well-defined conditions with particles collected from the river [36].

Using the numbers of surface groups for each of the components, an attempt can be made to transform the bulk composition of particles, in terms of dry weight, into numbers of surface groups per gram of particles. Two different examples of the results are shown in Figure 1, in order to illustrate possible distributions of surface groups. Some examples of the composition of particles in terms of dry weight are shown in Figure 1a, including two examples from rivers (average composition over a number of samplings at various times) and three examples from a lake at different times. The fraction of silicates (clay minerals, silicate in diatoms, etc.), which is not represented in this figure, would complete the composition to 1000 mg/g.

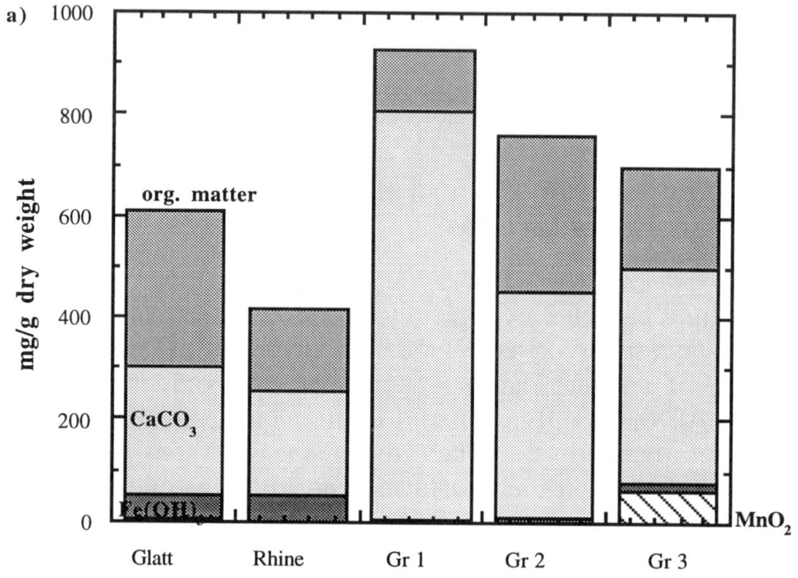

Figure 1. Composition of Suspended Solids and Estimated Number of Surface Groups. The suspended solids are from River Glatt, a small river with large inputs from sewage treatment plants and from an upstream lake, and from the River Rhine downstream of Basel (Switzerland) (in these two cases average composition over a number of samples). The three other samples are from Lake Greifen at different times of the year; Gr1 is representative of particles from the epilimnion in summer, Gr2 of the epilimnion in fall and Gr3 of the hypolimnion in winter. a. composition of the particles in terms of dry weight; b. estimated number of surface groups, using adsorption capacities given in Table 3, with amorphous $Fe(OH)_3$ and the value for binding of H^+ by the algae for organic material; c. estimated number of surface groups, using adsorption capacities given in Table 3, with α-FeOOH and the value for binding of $Cu^{2+}$by the algae for organic material.

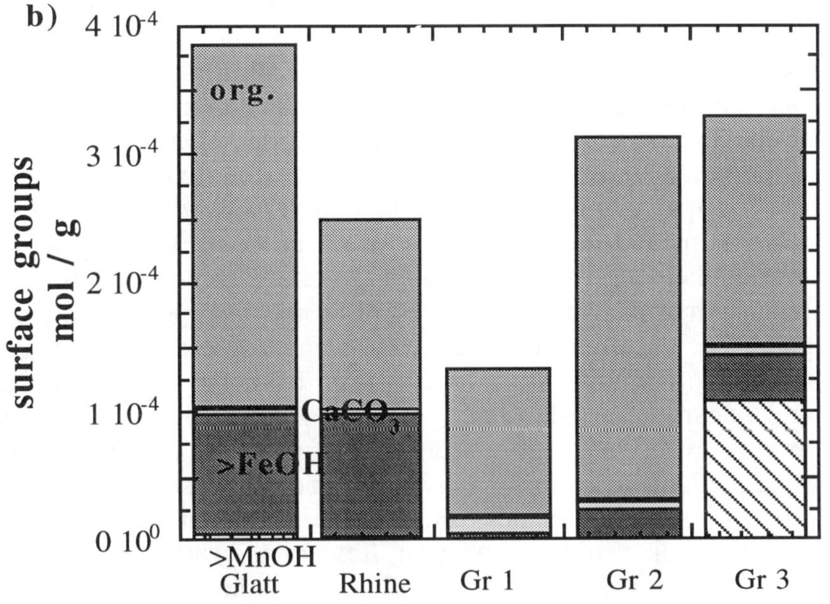

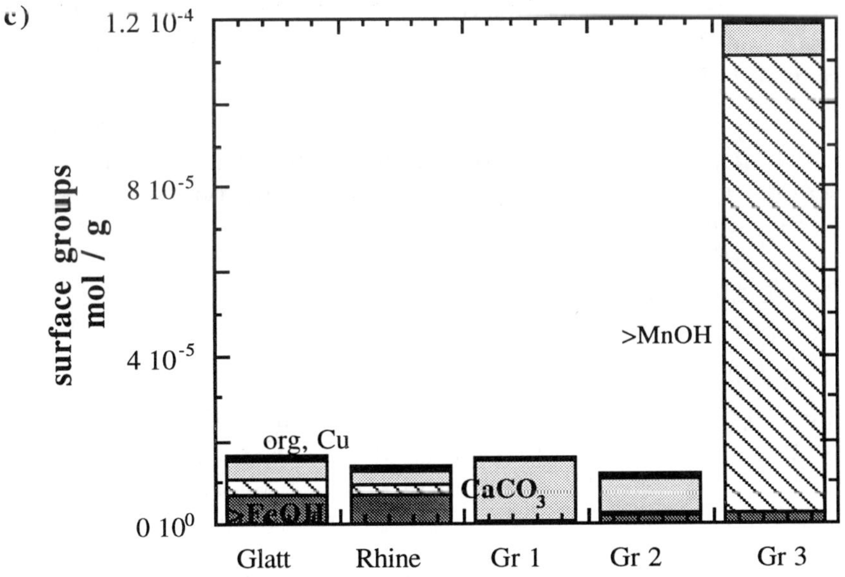

Figure 1. (Continued).

Estimates of the number of surface groups are presented in Figures 1b and 1c. In Figure 1b, the surface groups were estimated using the high value obtained with H^+ for the algae for organic material, and also a high number of surface groups on amorphous iron oxide; in Figure 1c only the specific groups for the adsorption of Cu were taken into account for organic material, and the estimate for goethite was taken as the number of surface groups on iron oxide. These figures confirm the importance of organic material and of iron oxides in providing surface groups for adsorption, both with respect to high numbers of surface groups and to high affinity for metal ions. Manganese oxides, because of their high surface areas and surface group density, may be important when their amount represents some few percents of the dry weight. The assumption of algae as being representative of organic material is certainly justified for particles from the lake, in which the organic material consists either of algae or of their debris. Organic debris may even have higher adsorption capacities than living algae, as more functional groups may become exposed upon partial decomposition. The particles from the rivers may contain other types of organic material; a high number of functional groups would also be obtained under the assumption of fulvic or humic acids, as the number of protonable groups for humic and fulvic acids are in the range $1 - 10 \times 10^{-3}$ mol/g [39]. The importance of iron oxides in providing a high number of surface groups, with high affinity for metal ions [19], has been demonstrated in a variety of environments [18]. Studies by Tessier *et al.* [40,41] have shown that the adsorption of metal ions in various environments with high amounts of iron oxides can be explained by the affinity of metal ions to iron oxides.

The clay minerals have not yet been considered in these calculations; they would provide an additional number of surface functional groups, depending on their size and types, as well as ion exchange sites.

4.4 Calculated Partition Coefficients

Stability constants of metal ions with surface groups of defined solids can also be used to evaluate the partition coefficients under certain assumptions. Calculations of the expected K_D for Zn and Pb were performed at given pH, using $Fe(OH)_3$, α-FeOOH, MnO_2, as well as calcite for Zn. The stability constants and number of surface groups used were from Dzombak and Morel [19] for amorphous $Fe(OH)_3$, from Balistrieri and Murray [27,29] for α-FeOOH and MnO_2, as well as from Müller and Sigg [42] for α-FeOOH, and from Zachara *et al.* [43] for calcite. The results obtained at pH 8 with α-FeOOH and with

MnO_2 were similar to the field results given in Table 1, whereas with amorphous $Fe(OH)_3$ the calculated K_D appeared to be much higher than the field results and with calcite they were lower.

The effects of dissolved ligands on partition coefficients can also be estimated from calculations including defined surface reactions. Some calculations for Cu are presented in Figure 2. They show the calculated partition coefficients, assuming amorphous $Fe(OH)_3$ as the solid phase, and either only the hydroxo complexes in solution, or in addition carbonate ($C_T = 2 \times 10^{-3}$ M), or as an example of an organic ligand, ethylenediamine (2×10^{-7} M). These calculations show how a ligand like ethylenediamine affects the pH dependence of the partition coefficient, resulting in a lower increase of the partition coefficient with pH than predicted in the absence of ligands other than OH^-, and possibly even in a decrease at high pH. The range of experimental (field and laboratory) data for Cu according to Tessier [18] is also shown. The experimental data, which are in a lower range than the calculated values with amorphous $Fe(OH)_3$, show a rather weak pH dependence, which may be attributed to the effects of natural organic ligands on Cu.

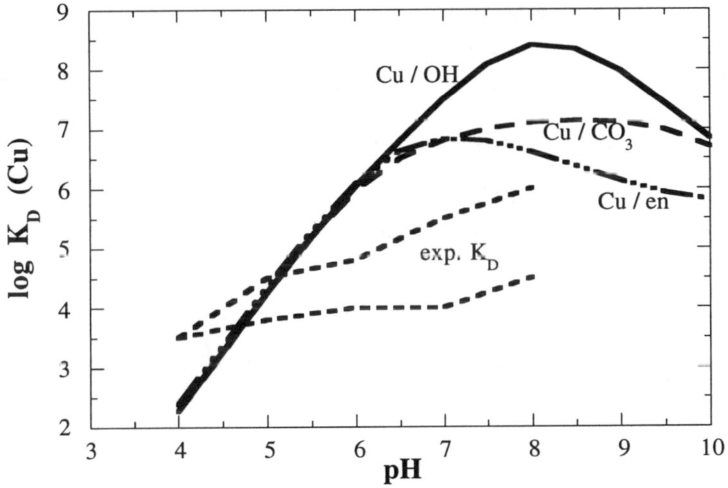

Figure 2. ***Calculated K_D for Cu ($Cu_T = 2 \times 10^{-8}$ M). The calculation was performed assuming amorphous Fe $(OH)_3$ as a solid phase (constants from Dzombak and Morel [19]), and hydroxo complexes (Cu/OH), or additionally carbonate complexes ($C_T = 2 \times 10^{-3}$ M, Cu/CO_3), or ethylenediamine complexes (en = 2×10^{-7} M, Cu/en). The dashed lines represent the range of experimental K_D for Cu (field and laboratory) according to Tessier [18].***

Realistic predictions of partition coefficients on the basis of the bulk composition of particles would necessitate better estimates of the number and affinity of actually active surface groups. It is very difficult to transform, e.g., a total iron oxide content, into a realistic number of surface sites. Furthermore, competitive effects between various surface groups, as well as competition between trace metal ions and major ions for adsorption sites should be evaluated.

5. ROLE OF PARTITION BETWEEN WATER AND SUSPENDED PARTICLES FOR THE FATE OF METAL IONS

The fraction of metal ions in solution depends on the concentration of particles in a given system. Variations in the concentrations of suspended solids are of special importance in rivers, in which the suspended solids concentrations strongly vary with water discharge [44]. The concentrations of particles may range in rivers from few mg/L to up to g/L. Especially some of the large rivers of the world carry high particle loads (e.g., Amazon, Yellow River, Ganges, etc. [44]). In some other rivers with higher ionic strength, the typical concentrations of suspended solids are lower, for example 5 - 50 mg/L in the Rhine [45]. In lakes, typical suspended particle concentrations are of the order of a few mg/L.

The relative fraction of metal ions in solution and in particles can be calculated from the partition coefficient and the suspended particle concentration:

$$\frac{c_W}{c_T} = \frac{1}{(1 + K_D \times S_w)} \tag{6}$$

where

c_W = concentration in solution (µg/L or mol/L)

c_T = total concentration (µg/L or mol/L)

K_D = partition coefficient (L/kg)

S_w = concentration of suspended solids (kg/L)

Figure 3 shows the fraction of metals in solution in function of the concentrations of suspended solids, for typical partition coefficients for the metals, in the range log K_D = 4 - 6. This figure illustrates that at low suspended solids concentrations, such as those typical of lakes, a large fraction of the metal ions is in solution, whereas at the high concentrations of suspended solids encountered in rivers the metal ions are mostly bound to suspended particles. In the case of rivers, with particle concentrations in the range 10 - 100 mg/L, the fraction in solution depends strongly on the partition coefficient.

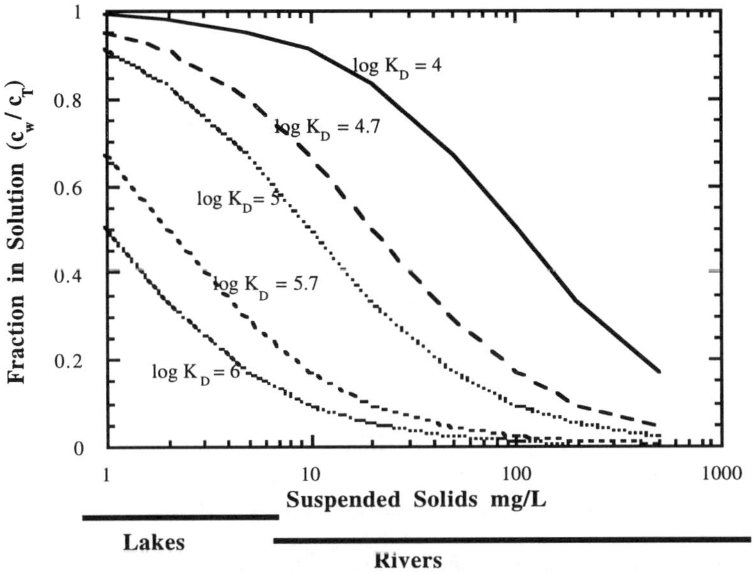

Figure 3. ***Fraction of Metals in Solution in Function of Suspended Solids for Various K_D; the Typical Range of Suspended Solids in Lakes and Rivers is Shown.***

The transport of metal ions primarily with the suspended solids is most relevant for large rivers with high suspended solids loads. For rivers with lower suspended solids concentrations, the metals in solution represent a significant part of the total metals. The latter case may be important in many metal polluted rivers, such as the Rhine. As an example, the fraction of dissolved Zn in the Rhine River (site Village-Neuf, below Basel, Table 1), was in most cases in the range 0.5-0.8 of the total Zn, with suspended solids ≈ 2 - 20 mg/L (Figure 4). For the direct interactions with the biota, the dissolved concentrations are of primary importance; the dissolved concentrations may be regarded as being buffered by the concentrations of suspended solids. High total concentrations of metals in rivers at high water discharges are usually mostly associated with the suspended particles; they do not usually mean that the dissolved concentrations also increase to the same extent.

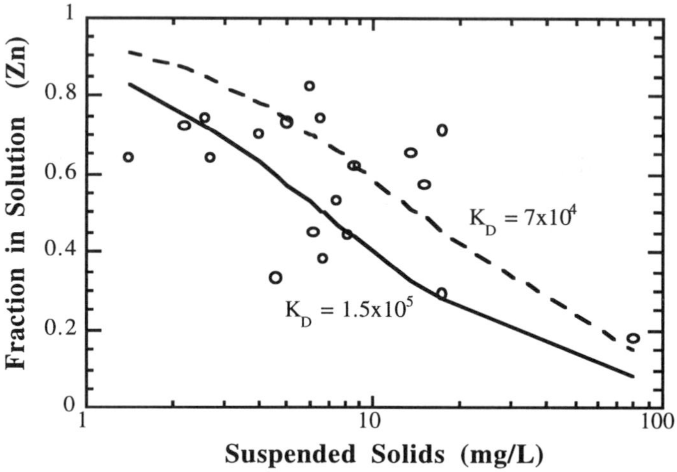

Figure 4. ***Fraction of Zn in Solution in Samples Taken at Various Times in the River Rhine in Function of Suspended Solids; the Typical Range of K_D is Shown.***

The concentrations of metals in lakes are mostly regulated by removal by sedimentation [2,3]. Although relatively low fractions of metal ions are present in suspended form in the water column of lakes (Figure 3), removal by sedimentation is usually an efficient removal mechanism, because of the high overall sedimentation fluxes, especially in eutrophic lakes. Low concentrations of metal ions in the water columns of lakes result from this efficient removal mechanism.

6. CONCLUSIONS

Field and laboratory data on partition of metals between solution and suspended particles indicate some general features. The pH dependence of the partition, as well as differences in affinities between metal ions, are especially well established. These tendencies are in general agreement with the partition being mostly determined by surface complexation reactions with various types of surface groups. The role of complexation of metal ions in solution as being in competition with binding to surfaces is, however, poorly understood. In most field investigations of partition of metals with suspended particles, complexation of the metals in solution has not been investigated. Differences in partition coefficients in various environments might be partly due to differences in the effects of ligands present in solution. In a simple case, competition between complexation in solution and at surfaces takes place; the adsorption of metal complexes may, however, play a role in some cases. It would be of interest to obtain field data sets on partition of metal ions that include a characterization of the speciation in solution.

Indirect evidence points in many cases to the important role of iron and manganese oxides, and of organic material (biota and biological debris), as being relevant components for the adsorption of metal ions. It remains, however, a difficult problem to understand the surface chemical properties of the natural assemblage of suspended particles, especially to relate the bulk composition of suspended particles to their surface chemical properties. The number and types of surface groups depend on the structures of the solids actually present, on the surface areas, and on the presence of certain components as coatings on certain particles (e.g., iron oxides). Further progress could be made by studying the surface properties of various fractions of natural particles, and by considering competition effects of various metal ions.

REFERENCES

1. Sunda, W.G., "Trace Metal Interactions with Marine Phytoplankton," *Biol. Oceanogr.* 6:411-442 (1988/89).

2. Sigg, L., "Metal Transfer Mechanisms in Lakes; The Role of Settling Particles," in *Chemical Processes in Lakes*, W. Stumm, Ed., pp. 283-310. Wiley Interscience (New York, NY 1985).

3. Sigg, L., "Surface Chemical Aspects of the Distribution and Fate of Metal Ions in Lakes," in *Aquatic Surface Chemistry*, W. Stumm, Ed., pp. 319-349. Wiley Interscience (New York, NY 1987).

4. Whitfield, M. and D.R. Turner, "The Role of Particles in Regulating the Composition of Seawater," in *Aquatic Surface Chemistry*, W. Stumm, Ed., pp. 457-493. Wiley Interscience (New York, NY 1987).

5. Stumm, W., *Chemistry of the Solid-Water Interface,* John Wiley (New York, NY 1992).

6. Schindler, P.W., "Surface Complexation," in *Metal Ions in Biological Systems*, H. Sigel, Ed., Vol. 18, Marcel Dekker (New York, NY 1984).

7. Sigg, L. Unpublished results, EAWAG, Dübendorf.

8. Internationale Kommission zum Schutze des Rheins (IKSR). Integrierte Bestandsaufnahme der aktuellen Qualität der Teilbereiche Wasser, Schwebstoff/Sediment und Organismen im Rhein, Report IKSR, (Koblenz, 1990).

9. Huynh-Ngoc, L., N. Whitehead and B. Oregioni, "Distribution of Zn in a Northwestern Mediterranean River," *Water, Air, Soil Pollut.* 34:315 -324 (1987).

10. Huynh-Ngoc, L., N. Whitehead and B. Oregioni, "Low Levels of Copper and Lead in a Highly Industrialized River," *Toxicol. Environ. Chem.* 17:223-236 (1988).

11. Elbaz-Poulichet, F., P. Holliger, W. Wen Huang and J.-M. Martin, "Lead Cycling in Estuaries, Illustrated by the Gironde Estuary, France," *Nature* 308:409-414 (1984).

12. Valenta, P., E.K. Duursma, A.G.A. Merks, H. Rützel and H. W. Nürnberg, "Distribution of Cd, Pb and Cu Between the Dissolved and Particulate Phase in the Eastern Scheldt and Western Scheldt Estuary," *Sci. Tot. Environ.* 53:41-76 (1986).

13. Lum, K.R., K.L.E. Kaiser and C. Jaskot, "Distribution and Fluxes of Metals in the St. Lawrence River from the Outflow of Lake Ontario to Québec City," *Aquat. Sci.* 53:1-19 (1991).

14. Shiller, A. M. and E. Boyle, "Dissolved Zinc in Rivers," *Nature* 317:49-52 (1985).

15. Balls, P.W., "The Partition of Trace Metals Between Dissolved and Particulate Phases in European Coastal Waters: A Compilation of Field Data and Comparison with Laboratory Studies," *Neth. J. Sea Res.* 23:7-14 (1989).

16. Benoit, G., S.D. Oktay-Marshall, A. Cantu, E.M. Hood, C.H. Coleman, M.O. Corapcioglu and P.H. Santschi, "Partitioning of Cu, Pb, Ag, Zn, Fe, Al and Mn Between Filter-retained Particles, Colloids and Solution in Six Texas Estuaries," *Mar. Chem.* 45:307-336 (1994).

17. Martin, J.M., D. Guan, M.F. Elbaz-Poulichet, A.J. Thomas and V.V. Gordeev, "Preliminary Assessment of the Distributions of Some Trace Elements (As, Cd, Cu, Fe, Ni, Pb and Zn) in a Pristine Aquatic Environment: The Lena River Estuary (Russia)," *Mar. Chem.* 43:185-199 (1993).

18. Tessier, A., "Sorption of Trace Elements on Natural Particles in Oxic Environments," in *Environmental Particles, Vol.1*, J. Buffle and H.P. van Leeuwen, Eds., pp. 425-453. Lewis Publishers (Boca Raton, FL 1992).

19. Dzombak, D. and F.M.M. Morel, *Surface Complexation Modeling; Hydrous Ferric Oxide* John Wiley (New York, NY 1990).

20. Schindler, P.W. and W. Stumm, "The Surface Chemistry of Oxides, Hydroxides and Oxide Minerals," in *Aquatic Surface Chemistry*, W. Stumm, Ed., pp.83-110. Wiley Interscience (New York, NY 1987).

21. Martin, J.-M., M. Dai and G. Cauwet, "Significance of Colloids in the Biogeochemical Cycling of Organic Carbon and Trace Metals in the Venice Lagoon (Italy)," *Limnol. Oceanogr.* 40:119-131 (1995).

22. Benoit, G., "Evidence of the Particle Concentration Effect for Lead and Other Metals in Fresh Waters Based on Ultraclean Technique Analyses," *Geochim. Cosmochim. Acta* 59:2677-2687 (1995).

23. Morel, F.M.M. and P.M. Gschwend, "The Role of Colloids in the Partitioning of Solutes in Natural Waters," in *Aquatic Surface Chemistry*, W. Stumm, Ed., pp. 405-422. Wiley Interscience (New York, NY 1987).

24. Honeyman, B.D. and P.H. Santschi, "The Role of Particles and Colloids in the Transport of Radionuclides and Trace Metals in the Oceans," in *Environmental Particles, Vol.1*, J. Buffle and H.P. van Leeuwen, Eds., pp. 379-423. Lewis Publishers (Boca Raton, FL 1992).

25. Mouvet , C. and A.C.M. Bourg, "Speciation (Including Adsorbed Species) of Cu, Pb, Ni And Zn in The Meuse River," *Wat. Res.* 17:641-649 (1983).

26. Nyffeler, U.P., Y.-H. Li and P.H. Santschi, "A Kinetic Approach to Describe Trace-Element Distribution Between Particles and Solution in Natural Aquatic Systems," *Geochim. Cosmochim. Acta* 48:1513-1522 (1984).

27. Balistrieri, L.S. and J.W. Murray, "Metal-Solids Interactions in the Marine Environment: Estimating Apparent Equilibrium Binding Constants," *Geochim. Cosmochim. Acta* 47:1091-1098 (1983).

28. Balistrieri, L.S. and J.W. Murray, "Marine Scavenging: Trace Metal Adsorption by Interfacial Sediment from Manop Site H," *Geochim. Cosmochim. Acta* 48:921-929 (1984).

29. Balistrieri, L.S. and J.W. Murray, "The Surface Chemistry of Sediments from the Panama Basin: The Influence of Mn Oxides on Metal Adsorption," *Geochim. Cosmochim. Acta* 50:2235-2243 (1986).

30. Müller, B. and L. Sigg, "Interaction of Trace Metals with Natural Particle Surfaces: Comparison Between Adsorption Experiments and Field Measurements," *Aquatic Sci.* 52:75-92 (1990).

31. Bruland, K.W., J.R. Donat and D.A. Hutchins, "Interactive Influences of Bioactive Trace Metals on Biological Production in Oceanic Waters," *Limnol. Oceanogr.* 36:1555-1577 (1991).

32. Kiefer, E.K.E., "Die Bindung von Kupfer an die Algenarten *Cyclotella cryptica* (Bacillariophyceae) und *Chlamydomonas reinhardtii* (Chlorophycae)," Ph.D. Thesis, Swiss Federal Institute of Technology, Zurich Nr. 10786 (1994).

33. Knauer, K., R. Behra and L. Sigg, "Effects of Free Cu^{2+} and Zn^{2+} Ions on Growth and Metal Accumulation in Freshwater Algae," *Environ. Toxicol. Chem.* 16:220-229 (1997).

34. Möhl, W., H. Motschi and A. Schweiger, "Magnetic Resonance Studies of Cu (II) Adsorbed on the Surface of the Bacterium Klebsiella Pneumonia," *Langmuir* 4:580 (1988).

35. Crist, R.H., K. Oberholser, N. Shank and M. Nguyen, "Nature of Bonding Between Metallic Ions and Algal Cell Walls," *Env. Sci. Technol.* 15:1212-1217 (1981).

36. Müller, B., "Ueber die Adsorption von Metallionen an Oberflächen aquatischer Partikel," Ph.D. Thesis, Swiss Federal Institute of Technology, Zurich Nr. 8988 (1989).

37. Mason, Y., "Natural Manganese Oxides Associated with Metallogenium-like Particles as Scavengers of Metals in Lakes," Ph.D. Thesis, Swiss Federal Institute of Technology, Zurich Nr. 11393 (1995).

38. Zachara, J.M., C.E. Cowan and C.T. Resch, "Metal Cation/Anion Adsorption on Calcium Carbonate: Implications to Metal Ion Concentrations in Groundwater," in *Metals In Groundwater*, H.E. Allen, E.M. Perdue and D.S. Brown, Eds., pp. 37-71. Lewis Publishers (Chelsea, MI 1993).

39. Buffle, J. *Complexation Reactions In Aquatic Systems*, Ellis Horwood (Chichester, UK 1988).

40. Tessier, A., F. Rapin and R. Carignan, "Trace Metals in Oxic Lake Sediments: Possible Adsorption onto IronOxyhydroxides," *Geochim. Cosmochim. Acta* 49:183-194 (1985).

41. Tessier, A., R. Carignan, B. Dubreuil and F. Rapin, "Partitioning of Zinc Between the Water Column and the Oxic Sediments in Lakes," *Geochim. Cosmochim. Acta* 52:1511-1522 (1989).

42. Müller, B. and L. Sigg, "Adsorption of Lead (II) on the Goethite Surface: Voltammetric Evaluation of Surface Complexation Parameters," *J. Coll. Interf. Sci.* 148:517-532 (1992).

43. Zachara, J.M., C.E. Cowan and C.T. Resch, "Sorption of Divalent Metals on Calcite," *Geochim. Cosmochim. Acta* 55:1549-1562 (1991).

44. Eisma, D., *Suspended Matter in the Aquatic Environment*, Springer-Verlag (Berlin, Germany 1993).

45. Internationale Kommission zum Schutze des Rheins (IKSR). Zahlentafeln der physikalisch-chemischen Untersuchungen des Rheinwassers und des Schwebstoffs (Koblenz, Switzerland 1993).

THE REDOX-pH CHEMISTRY OF CHROMIUM IN WATER AND SEDIMENT

Ronald D. DeLaune, William H. Patrick, Jr. and Tingzong Guo
Wetland Biogeochemistry Institute
Louisiana State University
Baton Rouge, LA 70803

1. INTRODUCTION

Chromium is a widespread contaminant in the environment, primarily as the result of increasing urbanization and industrial activities. Focusing on toxicity and exposure potential, the U.S. Environmental Protection Agency designated Cr and its compounds as one of 17 chemicals posing the greatest threats to human health.

Chromium content varies widely with the type and nature of soil and sediments. Sandy sediment contains the lowest chromium content. Highest chromium content is associated with soil and sediment with fine grain size [1]. Mean chromium content of soil in the United States and Western Europe are reported to be 53 mg/kg and 56 mg/kg, respectively [2]. Chromium is an essential nutrient for animal and plant metabolism. When its concentration reaches 0.1 mg/g body weight chromium can ultimately become lethal [3,4].

Chromium can exist in several oxidation states from Cr(0), the metallic form, to Cr(VI). The most stable oxidation states of chromium in the environment are Cr(III) and Cr(VI) [5]. Besides the elemental metallic form, which is extensively used in alloys, chromium has three important valence forms:

Chromous compounds, CrX_2, containing divalent chromium Cr (II).

Chromic compounds, CrX_3, containing trivalent chromium Cr(III).

Chromate and dichromate compounds, CrO_4^{2-} and $Cr_2O_7^{2-}$, containing tetravalent chromium Cr(VI).

The trivalent chromic Cr(III) and the tetravalent dichromate Cr(VI) are the most important forms in the environmental chemistry of waters and sediments. The presence of chromium Cr(VI) is of particular importance because in this oxidation state Cr is water soluble and extremely toxic. They are also the only forms that undergo valence changes in the Eh-pH ranges encountered in natural systems. The solubility and potential toxicity of the chromium that enters into aquatic or sediment columns is governed to a large extent by oxidation-reduction reactions.

Environmental concerns with soluble and particulate forms of Cr(VI) in soils/sediment and water has created a need to understand the solubility and valence stability of Cr(VI) in soils and sediment that are highly enriched with Cr [6]. Trivalent Cr is only moderately soluble in soils with low pH [7,8]. The reduction of Cr(VI) to Cr(III) in soils will minimize migration of Cr in soils and sediment [6]. Accordingly, a better understanding of means to reduce soluble Cr(VI) to less soluble Cr(III) is needed for use in remediation and minimizing the environment hazard of chromium in aquatic systems.

Significant levels of chromium can enter aquatic or wetland environments. DeLaune *et al.* [9] report chromium levels of several thousand μg/g in the sediment profile of a stream in a Louisiana wetland receiving chromium input as a result of point source industry discharge (Figure 1). Reduction or elimination of chromium discharge by industry has reduced the levels of chromium in the sediment columns as shown by the near background levels in the surface of the profile. Sedimentation can remove or reduce the interaction of chromium with the stream water columns, thereby reducing the contamination of downstream ecosystems.

In addition to the oxidation status of the chromium ions, a variety of soil/sediment biogeochemical processes such as redox reactions, precipitation, sorption and complexation to organic ligands can determine the fate of chromium entering a wetland environment.

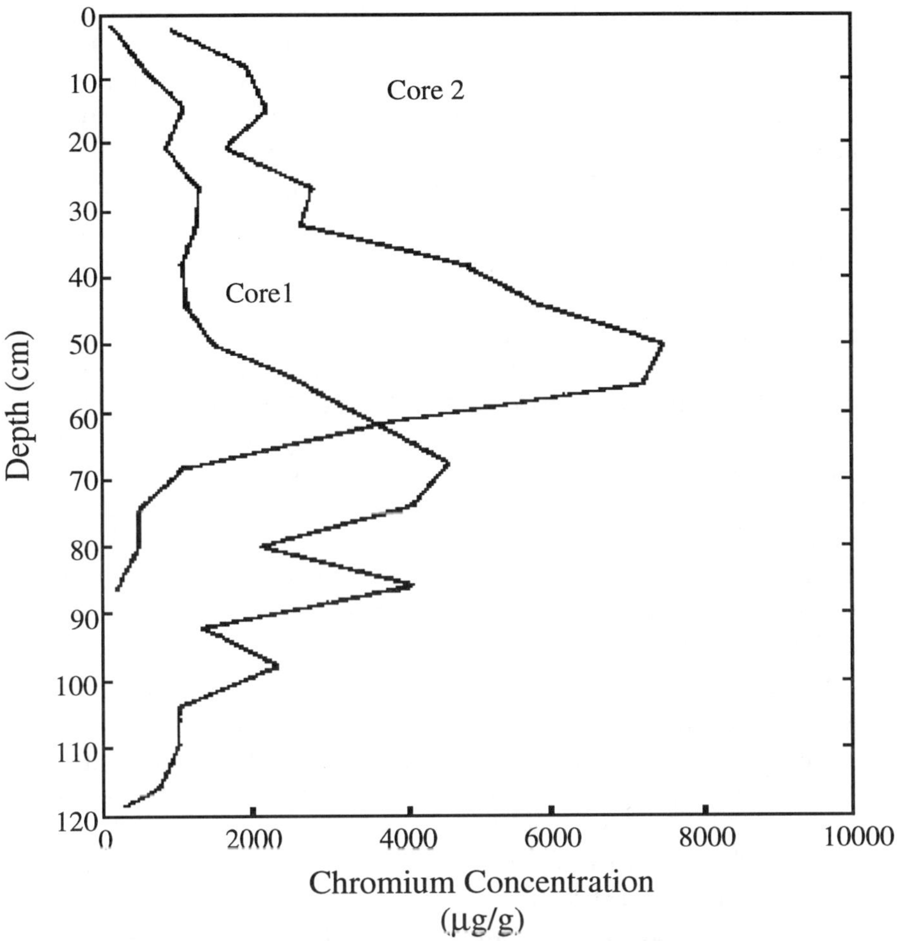

Figure 1. ***Profile Distribution of Chromium in Bottom Sediment of a Louisiana Stream Which has Received Point Source Industry Input.***

2. REDUCTION OF CHROMIUM AND OTHER ELECTRON ACCEPTORS IN SEDIMENT

The distribution between Cr(VI) and Cr(III) is regulated by an oxidation-reduction reaction in sediment. The redox transformation of Cr(VI) to Cr(III) or vice-versa can take place in the presence of other redox couples which can provide or accept three electrons. In an aquatic system the main redox couples are: O_2/H_2O, Mn(IV)/Mn(II), NO_3^-/NO_2^-, Fe(III)/Fe(II), SO_4^{2-}/S^{2-} and CO_2/CH_4 [5].

Oxygen is the terminal electron acceptor in aerobic systems and is reduced while organic electron donors are being oxidized. This reduction of O_2 to H_2O is carried out by true aerobic microorganisms and CO_2 is evolved as a waste product [10]. Therefore, a supply of oxidizable organic compounds, as well as a supply of O_2 and some means of removing CO_2 produced are indispensable for aerobic respiration to occur.

When the oxygen in the soil is consumed, the obligate aerobics can no longer function and the microbial community shifts to facultative anaerobic bacteria. When anoxic conditions persist, obligate anaerobes also proliferate. As O_2 is depleted, NO_3^- will be used as the electron acceptor. Many microorganisms can utilize NO_3^- as a terminal hydrogen acceptor instead of O_2, a process called NO_3^- respiration or dissimilatory NO_3^- reduction. Nitrate reduction to N_2 occurs at Eh values of 200 to 300 mV [11]. However, obligate anaerobic such as *Clostridium* can be responsible for dissimilatory NO_3^- reduction to NH_4^+ [12]. Significant correlations are observed between soil organic matter decomposition (expressed as CO_2 production) and NO_3^- respiration to N_2 indicating the use of NO_3^- as an electron acceptor [13].

Manganese is reduced following nitrate reduction. As the demand for electron acceptors increases, facultative anaerobes utilize manganic compounds as electron acceptors during their metabolic activity while using either endogenous or externally supplied substrates as electron donors [14-16]. The oxidized Mn compounds buffer Eh values in the range of +200 to +300 mV, until all of the bioreducible Mn has been converted to the Mn(II) form.

Ferric iron reductions sequentially follow manganese reduction; direct biological reduction is the most significant process in Fe(III) reduction. The Fe(III) compounds buffer the Eh of the system in the range of -100 to +100 mV [17]. For many mineral soils, the capacity of the Fe system is usually much greater than the Mn system. The ferric oxyhydroxide reduction occurs at the Eh values of approximately +150 mV.

Sulfate also serves as an electron acceptor. In anaerobic soils and sediments, two genera of strict anaerobes, i.e., *Desulfovibrio* and *Desulfotomaculum* are capable of utilizing SO_4^{2-}, as terminal electron acceptor for catabolic oxidation processes. This process involves the oxidation of substance C (electron donor) and transfer of electrons to SO_4^{2-} electron acceptor), a process termed as dissimilatory SO_4^{2-} reduction or SO_4^{2-} respiration. This mechanism results in a net release of free energy which is utilized for microbial cell synthesis. Only a small amount of reduced sulfur is assimilated by the organism and virtually all is released into the external environment as sulfide. The oxidation of organic matter during SO_4^{2-} reduction does not proceed beyond the level of acetate [18]. The SO_4^{2-} reduction occurs when the redox potential drops below -100 mV.

Methane is produced following sulfate reductions. Strict anaerobes which obtain their energy from coupling the oxidation of a limited range of substrates to the reduction of CO_2 to CH_4 are called methanogenic bacteria.

Oxidation-reduction processes also play a role in As and Se chemistry [19,20]. For the pH and redox conditions in most soils and sediments, arsenic (As) exists as an oxyanion in the arsenate [$H_2AsO_4^-$, and $HAsO_4^{2-}$; As(V)] or arsenite [H_3AsO_3; As(III)] form. Monomethylarsenic acid (MMAA) and dimethylarsenic acid (DMAA) are important organo-arsenic chemical species. Selenium (Se) occurs as selenate [SeO_4^{2-}; Se(VI)], selenite [$HSeO_3^-$ and SeO_3^{2-}; Se(VI)], elemental selenium [Se(0)], selenide (Se^{2-}) and organic selenides of which dimethyl selenide (DMSe) is the most important. Researchers are just now beginning to understand the geochemically important forms, and the quantity and kinetics of transfer among these different forms of As and Se. At higher soil redox levels (+200 to +500 mV), As(V) is the predominant As species present. The reduction of As(V) to As(III) occurs at redox levels corresponding within the nitrate reducing zone of soils characterized by a soil redox level of approximately +200 to +300 mV. When Fe reduction starts and soil redox levels drop below +50 mV, selenite is reduced to elemental Se or metal selenides. Reoxidation of reduced As and Se species were found to occur at similar soil redox levels as those identified for reduction reactions. Both microbial mediated and chemical oxidation processes are thought to be involved in the oxidation of As and Se species, but no real mechanism has been described.

Masscheleyn *et al.* [21] determined the rate of Cr(VI) reduction as affected by soil redox status. Soil suspensions were equilibrated under controlled redox conditions (-200, 0 +200 and +500 mV). After 14 days of equilibration, Cr(VI) was added to the suspensions at a rate of 50 mg of Cr (kg of dry soil)$^{-1}$. Samples were analyzed with time for soluble Cr(VI) and total soluble Cr. The amount of Cr removed (reduced) was calculated from the difference in the concentration of Cr added and the concentration remaining in solution after a certain reaction period.

Figure 2 shows the critical soil redox values for Cr(VI) reduction. Hexavalent chromium reduction occurred beginning at approximately the same redox levels as nitrate reduction. Water soluble Cr(VI) decreased in a linear relationship between 500 and 300 mV. As for nitrate, Cr(VI) reduction was complete at 300 mV. Figure 3 depicts the relation of Cr(VI) reduction in soil as compared to other redox couples.

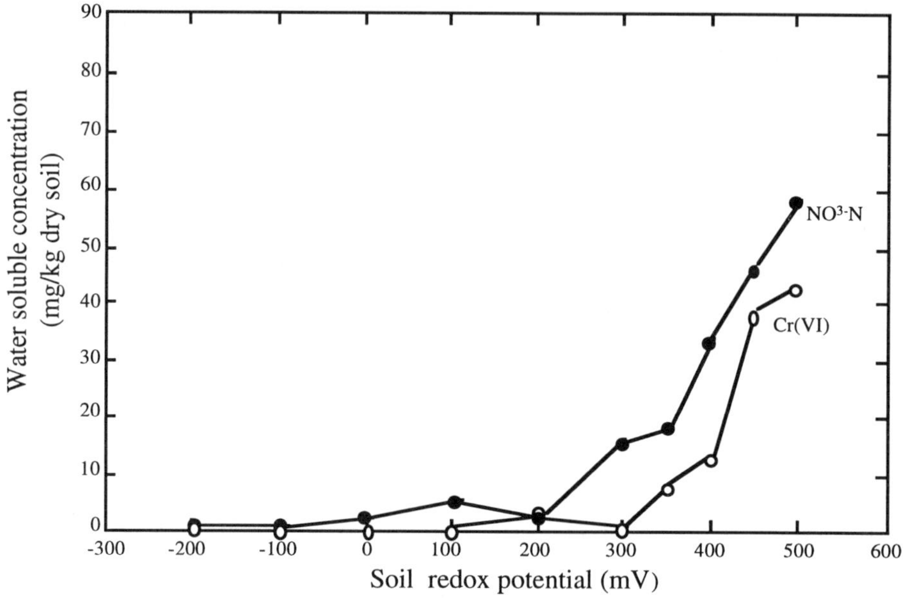

Figure 2. Critical Soil Redox Potential for Nitrate and Chromate Reduction.

3. CHROMIUM OXIDATION

Dissolved oxygen can oxidize Cr(III) to Cr(VI), however, the oxidation is slow at normal temperature in aqueous environments [22,23]. Due to the slow oxidation, Cr(III) is involved in the faster sorption and precipitation reactions [5]. Similar results were obtained by Masscheleyn *et al.* [21] investigating Cr(III) oxidation as affected by soil redox potential. In this reported soil study, Cr(III) was added at a rate of 500 mg of Cr (kg of dry soil)$^{-1}$ to a reduced (-200 mV) soil suspension. After 14 days of equilibration, a soil suspension aliquot was withdrawn and analyzed for soluble Cr(VI). The suspension was oxidized stepwise, with 100 mV increments, up to +500 mV, and dissolved Cr species were determined after a 4-day equilibration period at the preselected soil redox level. The stepwise oxidation of the soil suspension containing Cr(III) did not result in the production of Cr(VI).

Results showed that the Cr(VI) reduction dominated over the Cr oxidation reaction. The lack of Cr(III) oxidation in the soil suspension was attributed to precipitation (as $Cr(OH)_3$ or Cr_2O_3) or sorption in the soil. Cr(III) sorption-precipitation reactions were too rapid to measure, being complete in less than one minute

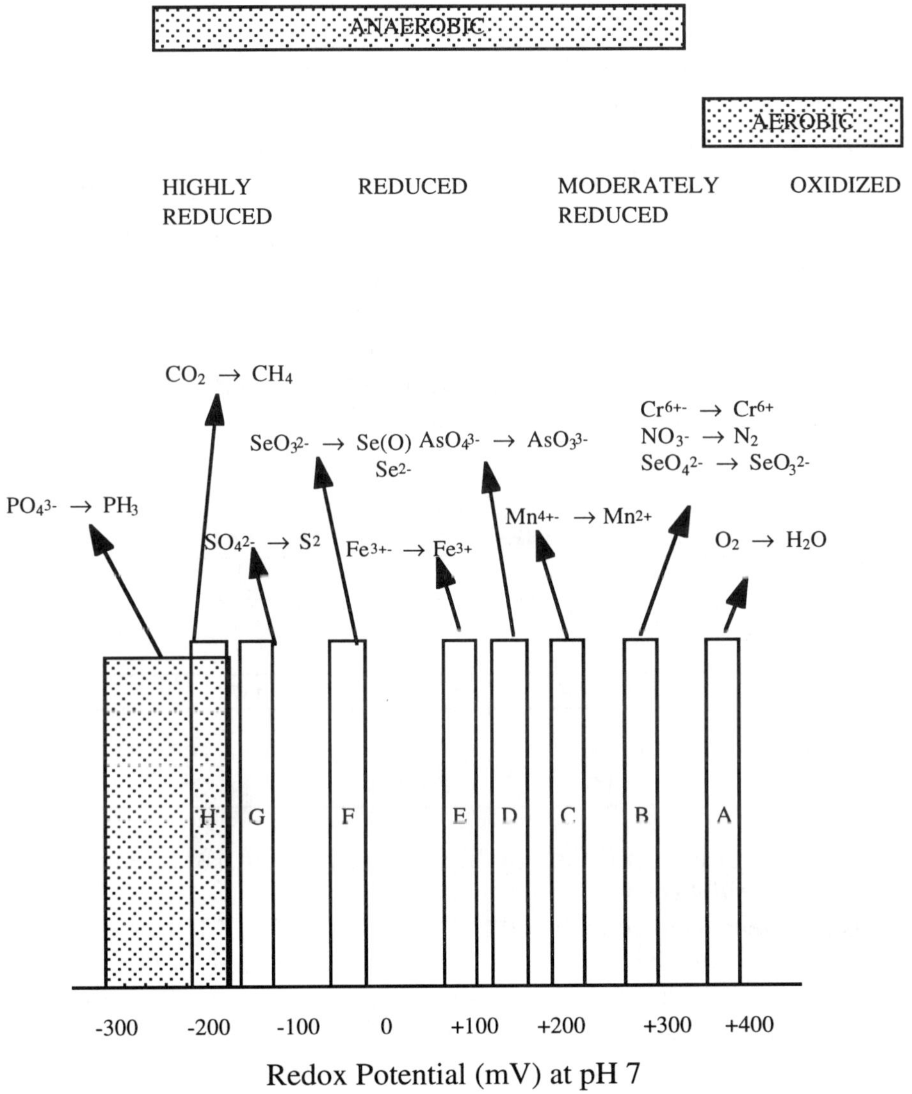

Figure 3. ***Critical Redox Potential for Transformations of Various Redox Couples in Sediment. A; [10], B; [11, 19, 20, 24], C; [16], D; [19, 25], E; [17, 26], F; [20], G; [27], H; [21], Shaded area; unknown.***

Bartlett and James [28] reported that trivalent chromium readily oxidized to the hexavalent form under some conditions prevalent in field soils. The major reaction by which Cr(III) is oxidized to the Cr(VI) form is chemical oxidation by manganese dioxide (MnO_2). Since oxygen (O_2) and nitrate (NO_3^-) are higher on the redox scale than MnO_2, it is unusual that MnO_2 is more effective than either of these better oxidants in oxidizing Cr(III). The reason is the capacity of MnO_2 to chemically oxidize Cr(III) is rapid while NO_3^- and O_2 are largely involved in slower biological oxidation reactions.

The oxidation of Cr(III) to Cr(VI) has also been reported in water column overlying reduced sediment [21]. Adsorption/precipitation reaction was slow enough to deter increase in Cr(VI) in the water column spiked with 1 and 10 mg L^{-1} Cr(III). In this study the majority of the Cr(III) initially added to the water column precipitated, but some remained in solution as Cr(III)-organic ligand complex. The observed oxidation was rapid with 12 and 57% of the 1 and 10 mg L^{-1} added Cr(III) being oxidized within 24 hrs. After one day, the Cr oxidation rate was exceeded by Cr(VI) removed (adsorptive difference). Table 1 depicts the quantities of Cr(VI) measured with time.

Table 1. Removal of chromium from floodwater of soil-water core spiked with Cr(III) [21].

	Soluble Cr (mg L^{-1})			
	1 mg Cr(III) L^{-1} added		10 mg Cr(III) L^{-1} added	
Days	Total Cr	Cr(VI)	Total Cr	Cr(VI)
0.25	0.54	0.12	5.5	4.0
3.00	0.25	0.07	6.0	3.6
6.00	0.19	0.05	4.0	2.2

The water column study by Masscheleyn *et al.* [21] demonstrated that Cr(III) oxidation to Cr(VI) can occur in the floodwater. However, it was not clear what contributed to Cr(III) oxidation. It was suggested that the relatively rapid rate of oxidation was coupled to the reduction dissolution of either manganese and/or iron oxide in the floodwater. However, additional studies are warranted on Cr(III) oxidation in both floodwater and sediment, especially in view of the fact that Cr(III) oxidation by oxygen is slow at earth surface temperatures.

4. CHROMIUM ADSORPTION

The solubility of chromium is strongly dependent upon its oxidation state, hexavalent chromium being more soluble than Cr(III). Cr(III), like other cationic metals, is rapidly adsorbed by soil Fe and Mn oxides and clay minerals [29,23,30]. Generally, added chromium is sorbed within 24 hrs by iron oxides and clay minerals in soil. Cr(III) adsorption increases with pH [5].

Soil redox status did not affect the capacity of the soil to retain Cr(III) [21]. Increase in the Cr(III) content of soil from 0 to 1500 mg Cr kg^{-1} did not affect the amount of water soluble Cr in soil suspension maintained at +500 mV, +200 mV, 0 mV and -200 mV.

Cr(VI), even though not as strongly as Cr(III), is also adsorbed by soil Mn, Al and Fe oxides and clay and organic colloids [31,22]. Adsorption of hexavalent chromium is reported to be a surface reaction between aqueous chromates and hydroxyl-specific surface sites. Each Cr(VI) surface complex covers 3-4 hydroxyl surface sites [32,33]. The reaction is pH dependent with Cr(VI) adsorption being favored on adsorbents which are positively changed at low to neutral pH [5].

Competing anions have a drastic effect on Cr(VI) adsorption. Chromate sorption in the natural environment is highly influenced by electrostatic conditions imposed by common anions (Cl^-, NO_3^-, SO_4^{2-}, HCO_3^- etc.) which are bound to the soil or sediment surface. Accordingly, adsorption of Cr(VI) is minimal in soil [22].

In contrast to Cr(III), soil redox condition was shown to strongly influence sorption of Cr(VI) from solution [21]. Under oxidized soil and moderately reduced (+500 to +100 mV) soil, chromium behavior was dominated by Cr(VI) sorption and reduction of Cr(VI) to Cr(III). Under more reduced soil redox levels (< +100 mV) chromium chemistry and solubility was controlled by the chemical reduction of Cr(VI) by soluble ferrous iron.

The Freundlich parameters obtained by Masscheleyn *et al.* [21] for the non-linear removal of Cr(VI) are shown in Table 2.

As reported by Masscheleyn *et al.* [21] the exact interpretation of Cr(VI) adsorption is difficult to determine from the adsorption isotherm. Even though there was a definite soil redox effect on Cr(VI) removal, the three differences we have described above have controlled Cr(VI) removal.

Table 2. Freundlich parameters for the removal of Cr(VI) as affected by soil suspension redox pH condition.

redox	pH	Kd	n
-200	7.3	822	2.85
0	6.8	466	3.30
200	6.0	60	2.56
500	5.7	1.14	1.22

5. INFLUENCE OF pH ON CHROMIUM BEHAVIOR

Low pH favors rapid reduction of Cr(VI) to Cr(III) [29]. James and Bartlett [7] report the reduction of soluble Cr(VI) in unlimed (pH 5.3) and limed (pH 6.5) soils resulted in the formation of soluble organic complexes of Cr(III) if citric acid was added with the Cr(VI). Greater reduction of Cr(VI) occurred in the unlimed soils.

Soil pH can also affect the form of Cr (VI) adsorbed on soil colloids. Below pH 6.4, $HCrO_4^-$ is the dominant chromium form in aqueous solution. Above pH 6.4, $HCrO_4^-$ dissociates to CrO_4^{2-} [34]. For their adsorption studies, James and Bartlett [8] reported that increasing soil pH from 5.4 to 7.0 by liming decreased the amount of chromium removed as phosphate exchangeable form from 7.2 kg/ha to 2.6 kg/ha.

Cr(III) is generally insoluble except under very acid condition, below approximately pH 4.0, which occurs only in soils and waters where pyrite oxidation has occurred. For this reason soluble Cr(III) is usually restricted to acid mine drainage and acid sulfate soils.

6. CHROMIUM CHEMICAL FRACTIONS IN SEDIMENT

Kinetics and chemical principles were used in quantifying the effects of sediment redox (Eh) condition on the behavior of Cr in various chemical fractions in sediment collected from a produced water discharge site in coastal Louisiana. Sediment samples were incubated in the controlled Eh-pH microcosms. Sediment was sequentially extracted over a 60 day period using the method of Shannon and White [35] for determining metals in various chemical fractions (water soluble, exchangeable, bound to carbonates, bound to iron and manganese oxides, bound to insoluble organic and sulfides) and chemical inactive fraction (mineral residue). Figure 4 shows the distribution of Cr in the various fractions. The majority (approximately 80%) of the chromium was bound in two fractions (iron and manganese oxides, sulfide and organic bound). Under oxidizing conditions, the behavior of Cr was controlled by Fe(III) and Mn(IV) oxides. Under reducing conditions, the behavior of Cr was controlled by insoluble high molecular weight humic materials and by sulfides. When redox potential increased past 0 mV, the affinity between Cr and Fe(III) and Mn(IV) oxides decreased. When redox potential decreased, especially at the more reducing soil redox levels, the affinity between Cr and insoluble sulfides and humic substances increased. An increase in water soluble chromium was observed between 0 and -150 mV. Less than 0.3 percent of the total chromium content of the sediment was in the water soluble form. The redox level in which soluble chromium increased represents a window following iron reduction and before significant sulfide is formed in the sediment.

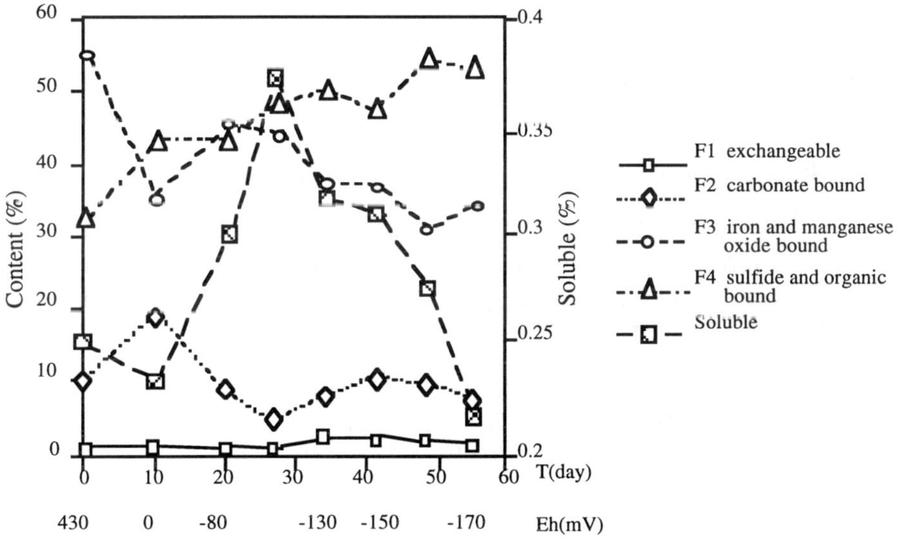

Figure 4. ***The Effect of Eh on the Percentage of Cr in the Various Chemical Active Forms.***

7. SUMMARY

Wetland soil/sediment redox and pH conditions play an important role in speciation and solubility of chromium in the aquatic environment. The mobility of chromium in the environment is determined by the competition among chromium oxidation state, adsorption/desorption mechanism and sediment redox-pH conditions. Chromium(VI) is reduced to chromium(III) at approximately +350 mV in soils and sediment. Reduced Cr(III) can be rapidly oxidized to the tetravalent chromate and dichromate forms by oxidized manganese compounds. Cr(III) is much less soluble in natural system than the hexavalent form and has a much lower toxicity. Chromium is less likely to be a problem in wetlands than in non-wetlands because the reducing conditions cause its reduction or conversion to the more insoluble Cr(III) form.

Acknowledgment: Work was supported by Louisiana State University Coastal Marine Institute funded by Minerals Management Service (Contract # 14-35-0001-30660).

REFERENCES

1. Salomons, W. and A.J. De Groot, "Pollution History of Trace Metals in Sediments, as Affected by the Rhine River," in *Environmental Biogeochemistry*, W.E. Krumbein, Ed., pp. 149-162. Ann Arbor Science (Ann Arbor, MI 1978).

2. Angelone, M. and C. Bini, "Trace Element Concentration in Soil and Plants of Western Europe," in *Biogeochemistry of Trace Elements*, D.C. Adriano, Ed., pp. 19-60. Lewis Publishers (Boca Raton, FL 1992).

3. Ajmal, M., A.A. Nomani and A. Ahmad, "Acute Toxicity of Chrome Electroplating Wastes to Microorganisms: Adsorption of Chromate and Chromium (VI) on a Mixture of Clay and Sand," *Wat. Air Soil Pollut.* 23:119-127 (1984).

4. Mertz, W., "Chromium as a Dietary Essential for Man," in *Trace Element Metabolism in Animals -. Vol. 2 Proc. Symp. on Trace Metabolism in Animals*, W.G. Hoekstra, J.W. Suttie, K.E. Ganther and W. Mertz, Eds., pp. 185-198. University Park Press (Madison, WI 1974).

5. Richard, F.C. and A.C.M. Bourg, "Aqueous Geochemistry of Chromium: A Review," *Wat. Res.* 25:807-816 (1991).

6. James, B.R., "Hexavalent Chromium Solubility and Reduction in Alkaline Soils Enriched with Chromite Ore Processing Residue," *J. Environ. Qual.* 23:227-233 (1994).

7. James, B.R. and R.J. Bartlett, "Behavior of Chromium in Soils: VI. Interactions Between Oxidation-Reduction and Organic Complexation," *J. Environ. Qual.* 12:173-176 (1983).

8. James, B.R. and R.J. Bartlett, "Behavior of Chromium in Soils: VII. Adsorption and Reduction of Hexavalent Forms," *J. Environ. Qual.* 12:177-181 (1983).

9. DeLaune, R.D., R.P. Gambrell, J.H. Whitcomb, G.L. Jones, D.E. Leach and J. Wiesepape, "Sediment Analysis and Sedimentation in Bayou Trepagnier," Final Report, Louisiana Department of Environmental Quality (1987).

10. Reddy, K.R. and W.H. Patrick, Jr., "Effect of Alternate Aerobic and Anerobic Conditions on Redox Potential, Organic Matter Decomposition and Nitrogen Loss in a Flooded Soil," *Soil Biol. Biochem.* 7:87-94 (1975).

11. Patrick, Jr., W.H., "Nitrate Reduction Rates in a Submerged Soil as Affected by Redox Potential," *7th Int'l. Cong. of Soil Sci.*, Madison, WI. 2:494-500 (1960).

12. Casey, W.H. and J.M. Tiedje, "The Reduction of Nitrate to Ammonium in Soils," *J. Gen. Microbiol.* 119:217-223 (1980).

13. Reddy, K.R., P.S.C. Rao and R.E. Jessup, "The Effect of Carbon Mineralization on Denitrification Kinetics in Mineral and Organic Soils," *Soil Sci. Soc. Am. J.* 46:62-68 (1982).

14. Takai, Y. and T. Kamura, "The Mechanism of Reduction in Waterlogged Paddy Soil," *Folia Microbiologia* 11:304-313 (1966).

15. Trimble, R.B. and H.L. Ehrlich, "Bacteriology of Manganese Nodules. III. Introduction of MnO_2 by Two Strains of Nodule Bacteria," *Appl. Microbiol.* 16:695-702 (1968).

16. Turner, F.T. and W.H. Patrick, Jr., "Chemical Changes in Waterlogged Soils as a Result of Oxygen Depletion," *Soil Sci.* 4:53-65. (1968).

17. Patrick, Jr., W.H., "The Role of Inorganic Redox Systems in Controlling Reduction in Paddy Soils," in *Proc. Symp. on Paddy Soils*. Inst. of Soil Sci. Acad., Sinica, China, Ed., pp. 107-117. Springer-Verlag (New York, NY 1981).

18. Wake, L.V., R.K. Christopher, A.D. Rickard, J.E. Anderson and B.J. Ralph, "A Thermodynamic Assessment of Possible Substrates for Sulfate Reducing Bacteria," *Aust. J. Biol. Sci.* 30:155-172 (1977).

19. Masscheleyn, P.H., R.D. DeLaune and W.H. Patrick, Jr., "Effect of Redox Potential and pH on Arsenic Speciation and Solubility in a Contaminated Soil," *Environ. Sci. Technol.* 25:1414-1419 (1991).

20. Masscheleyn, P.H., R.D. DeLaune and W.H. Patrick, Jr., "Transformation of Selenium as Affected by Sediment Oxidation-Reduction Potential and pH," *Environ. Sci. Technol.* 24:91-96 (1990).

21. Masscheleyn, P.H., J.H. Pardue, R.D. DeLaune and W.H. Patrick, Jr., "Chromium Redox Chemistry in a Lower Mississippi Valley Bottomland Hardwood Wetland," *Environ. Sci. Technol.* 26:1217-1225 (1992).

22. Rai, D, J.M. Zachara, L.E. Eary, D.C. Girvin, D.A. Moore, C.T. Resch, B.M. Sass and R.L. Schmidt, "Geochemical Behavior of Chromium Species," *Interim Report EPRI EA-4544*, Electric Power Research Institute, Palo Alto, CA (1986).

23. Schroeder, D.C. and G.F. Lee, "Potential Transformations of Chromium in Natural Waters," *Wat. Air Soil Pollut.* 4:355-365 (1975).

24. Buresh R.J. and W.H. Patrick, Jr., "Nitrate Reduction to Ammonium and Organic Nitrogen in an Estuarine Sediment," *Soil Biol. Biochem.* 13:279-283 (1981).

25. Masscheleyn, P.H., R.D. DeLaune and W.H. Patrick, Jr., "Arsenic and Selenium Chemistry as Affected by Sediment Redox Potential and pH," *J. Environ. Qual.* 20:522-527 (1991).

26. Patrick, Jr., W.H., "Extractable Iron and Phosphorus in a Submerged Soil at Controlled Redox Potentials," *Trans. 8th Intl. Cong. Soil Sci.,* Bucharest, Romania IV66:605-609 (1964).

27. Connell, W.E. and W.H. Patrick, Jr., "Sulfate Reduction in Soil: Effects of Redox Potential and pH," *Science* 159:86-87 (1968).

28. Bartlett, R.J. and B. James, "Behavior of Chromium in Soils. III. Oxidation," *J. Environ. Qual.* 8:31-35 (1979).

29. Bartlett, R.J. and J.M. Kimble, "Behavior of Chromium in Soils. I. Trivalent Forms," *J. Environ. Qual.* 5:370-383 (1976).

30. Dreiss, S.J., "Chromium Migration Through Sludge Treated Soils," *Groundwater* 24:312-321 (1986).

31. MacNaughton, M.G., "Adsorption of Chromium (VI) at the Oxide-Water Interface," *Interim report July 1974-June 1975*, Environ. Chem. Div., Environics Directorate, Air Force Civil Eng. Center, Tyndal AFB, FL (1975).

32. Davis, J.A. and J.O. Leckie, "Surface Ionization and Complexation at the Oxide/water Interface. 3. Adsorption of Anions," *J. Colloid Interface Sci.* 74:32-43 (1980).

33. Benjamin, M.M. and N.S. Bloom, "Effect of Strong Binding Anionic Adsorbates on Adsorption of Trace Metals on Amorphous Iron Oxyhydroxide," in *Adsorption from Aqueous Solution*, P.H. Tewari, Ed., pp. 41-60. Plenum Press (New York, NY 1981).

34. Deltcombe, E., N. deZoubov and M. Pourbaix, "Chromium," in *Atlas of Electrochemical Equilibria in Aqueous Solutions*, M. Pourbaix, Ed., pp. 256-271. Pergamon Press (Oxford, UK 1966).

35. Shannon, R.D. and J.R. White, "The Selectivity of a Sequential Extraction Procedure for the Determination of Iron Oxyhydroxides and Iron Sulfides in Lake Sediments," *Biogeochem.* 14:193-208 (1991).